WIRELESS COMMUNICATION TECHNOLOGIES : NEW MULTIMEDIA SYSTEMS

Edited by

NORIHIKO MORINAGA
Osaka University

RYUJI KOHNO
Yokohama National University

SEIICHI SAMPEI
Osaka University

Kluwer Academic Publishers
Boston/Dordrecht/London

Distributors for North, Central and South America:
Kluwer Academic Publishers
101 Philip Drive
Assinippi Park
Norwell, Massachusetts 02061 USA
Telephone (781) 871-6600
Fax (781) 871-6528
E-Mail <kluwer@wkap.com>

Distributors for all other countries:
Kluwer Academic Publishers Group
Distribution Centre
Post Office Box 322
3300 AH Dordrecht, THE NETHERLANDS
Telephone 31 78 6392 392
Fax 31 78 6546 474
E-Mail services@wkap.nl>

 Electronic Services <http://www.wkap.nl>

Library of Congress Cataloging-in-Publication

Wireless communication technologies : new multimedia systems / edited by Norihiko Morinaga, Ryuji Kohno, Seiichi Sampei.
   p. cm. -- (The Kluwer international series in engineering and computer science ; SECS 564)
   Includes bibliographical references and index.
   ISBN 0-7923-7900-4 (alk. paper)
   1. Wireless communication systems. 2. Multimedia systems. I. Morinaga, Norihiko, 1939- II. Kohno, Ryuji, 1956- III. Sampei, Seiichi. IV. IEEE International Symposium on Personal, Indoor, and Mobile Radio Communications (10th : 1999 : Osaka, Japan) V. Series.

   TK5103.2. W5719 2000
   621.382--dc21

                                                                    00-031339

Contents

Preface

During 12-15 of September 1999, 10th International Symposium on Personal, Indoor and Mobile Radio Communications (PIMRC'99) was held in Osaka Japan, and it was really a successful symposium that accommodated more than 600 participants from more than 30 countries and regions. PIMRC is really well organized annual symposium for wireless multimedia communication systems, in which, various up-to-date topics are discussed in the invited talk, panel discussions and tutorial sessions.

One of the unique features of the PIMRC is that PIMRC is continuing to publish, from Kluwer Academic Publishers since 1997, a book that collects the hottest topics discussed in PIMRC. In PIMRC'97, Invited talks were summarized in "Wireless Communications –TDMA versus CDMA – (ISBN 0-7923-8005-3)," and it was published just before PIMRC'97. This book was also distributed to all the PIMRC'97 participants as a part of proceedings for the conference. In PIMRC'98, extended version of the invited papers were summarized in Wireless Multimedia Network Technologies (ISBN 0-7923-8633-7) and published in September 1999, which is almost the same timing for the PIMRC'99.

In the case of PIMRC'99, to produce more informative book, we have selected topics that attracted many PIMRC'99 participants during the conference, and invited prospective authors not only from the invited speakers but also from tutorial speakers, panel organizers, panelists, and some other excellent PIMRC'99 participants.

This book is divided into two parts; Part I for new technical trends in wireless multimedia communications, and Part II for trends in new wireless multimedia communication systems that will be serviced in early 2000s.

In Part I, we have selected seven key technologies that strongly drive developments of new wireless multimedia communication systems; wireless channel modeling, space-time coding, coding for wireless, OFDM, multiuser receiver, software radio, and, spatial and temporal communication theory.

In Chapter 1, Mr. Gregory D. Durgin and Prof. Theodore S. Rappaport of MPRG, Virginia Tech, USA, propose a new spatial channel modeling techniques that characterizes angle-of-arrival of multipath fading, including its application to several practical wireless communication channels.

In Chapter 2, Dr. Ayman F. Nguib and Dr. Rob Calderbank of AT & T Labs-Research, USA, provide overview of space-time coding techniques including its associated signal processing framework which is attracting many 3G system engineers.

In Chapter 3, Prof. Ezio Biglieri and Prof. Giorgio Taricco of Politecnico di Torino, Italy, and Prof. Giuseppe Caire of Institut Eurecom, Sophia Antipolis, France, discuss how to select coding schemes considering not only the nature of channels but also the applied techniques such as bit-interleaving, diversity and transmit power control.

In Chapter 4, Prof. Shinsuke Hara of Osaka University, Japan, addresses OFDM techniques especially synchronization issues specific to the OFDM systems.

In Chapter 5, Prof. Markku Juntti and Mr. Kri Hooli of University of Oulu, Finland, provides overview of the multiuser receivers for CDMA systems including its basic principle, its combination with the multipath and antenna combining techniques and its potential applications.

In Chapter 6, Dr. Shinichiro Haruyama of Advanced Telecommunication Laboratory of SONY Computer Science Laboratories, Inc., Japan, addresses overview of the development of software-defined radio technologies including current activity of the SDR forum.

In Chapter 7, Prof. Ryuji Kohno of Yokohama National University, Japan, who is also one of the editors for this book introduces a spatial and temporal communication theory based on adaptive antenna array, such as channel modeling, equalization and joint optimization of spatial and temporal signal processing in both transmitter and receiver.

In Part II, because various new wireless systems are currently being standardized, we have selected five topics for new wireless systems, i.e., Intelligent Transport System, wireless data communication systems, wireless Internet, digital TV broadcasting and IMT-2000.

In Chapter 8, Dr. Masayuki Fujise, Dr. Akihito Kato, Dr. Katsutoshi Sato and Dr. Hiroshi Harada of Communications Research Laboratory (CRL), Ministry of Posts and Telecommunications, Japan, present key technologies for Intelligent Transport Systems currently developed by CRL; inter-vehicle and road-vehicle communications, radio-on-fiber and software radio technologies.

In Chapter 9, Prof. Kaveh Pahlavan and Mr. Xinrong Li of Worcester Polytechnic Institute, USA, and, Dr. Mika Ylianttila and Prof. Matti Latva-aho of Universty of Oulu, Finland, present overview of the current status and future trends of wireless data communication systems such as wireless LAN, HomeRF and Bluetooth.

In Chapter 10, Dr. Li Fun Chang of AT & T Labs - Research, USA, provides overview of the networking and mobility aspects of the wireless core networks, e.g. mobile-IP based and EGPRS-based networks including basic concept of mobile IP for both IPv4 and IPv6.

In Chapter 11, Prof. Makoto Itami of Science University of Tokyo, Japan, addresses overview of the digital terrestrial TV broadcasting systems in EU, USA and Japan including their feature comparison.

In Chapter 12, Prof. Fumiyuki Adachi of Tohoku University, Japan, and Dr. Mamoru Sawahashi of NTT DoCoMo, Japan, discuss evolution of cellular phone systems from voice services to multimedia services, IMT-2000 standardization activities towards global 3G standard, and W-CDMA technologies including some advanced technologies such as interference cancellation and adaptive array antenna.

Because each chapter includes basic concept and technical trend in addition to the main topics, this book is suitable not only for the research engineers who are developing 3G systems but also the graduate course students who would like to know what is the cutting edge technologies, or managers in industries to understand technical trends of the wireless world.

We, as the editors of this book, appreciate all the authors for their cooperation in preparing for such up-to-date and informative contents.

Finally, the editors would like to appreciate those who helped us in editing final version of the manuscript of this book. Especially, we would like to express our sincere appreciation to Mr. Takumi Ito who spent a lot of time in making final electric manuscript in LaTeX format including file conversion from MsWord to LaTeX, and Mr. Tomoaki Yoshiki who helped in creating index files.

Norihiko Morinaga
Ryuji Kohno
Seiichi Sampei

This book is dedicated to those who volunteered to organize PIMRC'99.

I

NEW TECHNICAL TREND IN WIRELESS MULTIMEDIA COMMUNICATIONS

Chapter 1

SPATIAL CHANNEL MODELING FOR WIRELESS COMMUNICATIONS

Gregory D. Durgin
Mobile and Portable Radio Research Group
Bradley Department of Electrical and Computer Engineering Virginia Tech
gdurgin@vt.edu

Theodore S. Rappaport
Mobile and Portable Radio Research Group
Bradley Department of Electrical and Computer Engineering Virginia Tech
wireless@vt.edu

Abstract   This chapter presents a novel theoretical framework for relating the small-scale fading characteristics of a wireless channel to multipath angle-of-arrival. A method is presented for reducing a multipath channel with arbitrary spatial complexity to three shape factors that have simple, intuitive geometrical interpretations. Furthermore, these shape factors are shown to describe the statistics of received signal fluctuations in a fading multipath channel. Examples demonstrate how the shape factors may be applied to real-life problems in channel measurement, level-crossing rate and average fade duration calculations, and coherence distance estimation.

Keywords: angle-of-arrival, channel modeling, multipath, radio wave propagation, small-scale fading

1   INTRODUCTION

At the start of the new millennium, an exciting perspective is emerging in the field of wireless channel modeling. In the past, the wireless multipath channel was thought to be a harsh, unavoidable consequence of wireless communications. In recent years, new technology in hardware and channel coding have not just overcome the difficulties of communicating in a multipath channel – algorithms such as space-time coding actually use the unpredictable nature of the multipath channel to *enhance* the communications link [1].

One fact is inescapable: the development of new wireless systems requires that the channel be measured and modeled to an increasingly higher degree of detail. It no longer suffices to make oversimplifying assumptions about the spatial channel, such as omnidirectional multipath propagation and Rayleigh fading. Multiple antenna receivers cannot function properly if designed without an understanding of the spatio-temporal characteristics of the multipath channel.

Multipath propagation leads to two unpredictable types of behavior in the wireless channel. The first is *frequency selectivity* caused by multipath components arriving with different delays. The second is *spatial selectivity* caused by multipath components arriving from different directions in space. While frequency selectivity is a well-understood phenomenon, the problem of describing spatial selectivity, which results in *small-scale fading*, has traditionally been difficult for wireless engineers to model for emerging space-time applications. There is a need to relate basic small-scale fading characteristics to the spatial geometry of arriving multipath.

This chapter presents a theoretical framework for characterizing the angle-of-arrival of multipath power in a way that produces simple-but-powerful insight into the nature of small-scale fading. By emphasizing the parallel mathematical analysis used for frequency selectivity and spatial selectivity, we show that small-scale fading behavior may be described with only three geometrical angle-of-arrival parameters: angular spread, angular constriction, and azimuthal angle of maximum fading. These three *shape factors* relate to spatial selectivity much like RMS delay spread relates to frequency selectivity.

The rest of the chapter is broken into the following sections: Section 2 discusses basic concepts in modeling stochastic wireless channels. Section 3 defines the three basic shape factors – geometrical parameters that describe multipath angles-of-arrival. Several examples illustrating the shape factor concept are found in Section 4. Section 5 then presents practical problems in wireless channel modeling which are solved easily by using multipath shape factors. The chapter concludes with a final perspective on the work presented.

2   THE BASICS OF SMALL-SCALE CHANNEL MODELING

This section discusses the use of a baseband channel model to explore two types of *local area* behavior in the wireless channel: frequency selectivity and spatial selectivity. A local area is a region in space (typically about 20 wavelengths for microwave mobile receivers) over which the mean power level of the channel is undisturbed by large-scale scattering and shadowing.

2.1   RECEIVED COMPLEX VOLTAGE

As propagating waves impinge upon an antenna, they excite an oscillating voltage at the input terminals of the receiver. This voltage is a function of position, r, of the receiver antenna. For time-harmonic (narrowband) analysis, it suffices to describe the received voltage in the form of a complex phasor, $\tilde{h}(r)$, that is solely a function of position. The received radio frequency voltage, $\mathcal{H}(r, t)$, is related to the complex voltage (also called the baseband voltage) by the following relationship:

$$\mathcal{H}(r, t) = \text{Real} \left\{ \tilde{h}(r) \exp(j 2\pi f_c t) \right\} \tag{1.1}$$

where f_c is the radiation carrier frequency [2].

Since the phasor transform in Eqn (1.1) completely captures the time dependence, all subsequent analysis will focus on the baseband representation of complex voltage, $\tilde{h}(r)$. From complex voltage, it is possible to calculate any of the following pieces of information:

$$
\begin{array}{rcl}
\text{In-Phase Received Component} & : & \text{Real}\{\tilde{h}(r)\} \\
\text{Quadrature Received Component} & : & \text{Im}\{\tilde{h}(r)\} \\
\text{Voltage Envelope, } R(r) & : & |\tilde{h}(r)| \\
\text{Power (units of } Volts^2), P(r) & : & |\tilde{h}(r)|^2
\end{array}
$$

The voltage envelope, $R(r)$, and received power, $P(r)$, are particularly important for analysis since they govern the *signal-to-noise ratio* of the communications link and ultimately determine the instantaneous quality of the wireless channel along a local area.

2.2   FREQUENCY SELECTIVITY

For a fixed, single-antenna receiver operating in a static channel, the principle source of channel distortion for a received signal is dispersion induced by multipath propagation delays. This time-dispersive channel is characterized

by a complex, baseband *channel-impulse response*, $\tilde{h}(\tau)$, which measures received voltage as a function of time-delay, τ. If a waveform, $s(\tau)$, is transmitted on the carrier wave, then the received waveform is given by the convolution
of $s(\tau)$ with $\tilde{h}(\tau)$.

A useful measure of dispersion in a wireless channel is the RMS delay
spread, σ_τ. This value is calculated from the delay spectrum, $p(\tau)$, which
is defined to be the average value of $|\tilde{h}(\tau)|^2$ in a local area. Much of the litera-
ture also refers to this quantity as the mean power delay profile. The definition
for RMS delay spread follows as the second centered moment of the delay
spectrum:

$$\sigma_\tau^2 = \overline{\tau^2} - (\overline{\tau})^2, \quad \text{where } \overline{\tau^n} = \frac{\int\limits_{-\infty}^{+\infty} \tau^n p(\tau) d\tau}{\int\limits_{-\infty}^{+\infty} p(\tau) d\tau} \qquad (1.2)$$

As the impulse response, $\tilde{h}(\tau)$, becomes broader in delay, the RMS delay
spread increases.

Besides being a useful metric, RMS delay spread has an insightful mathematical property regarding the frequency domain representation of $\tilde{h}(\tau)$. If we
view the evolution of $H(f)$, the Fourier transform of $\tilde{h}(\tau)$, in the frequency
domain as a wide-sense stationary stochastic process, then its mean-squared
derivative is proportional to the delay spread:

$$\mathrm{E}\left\{ \left| \frac{d[H(f)\exp(-j2\pi f\overline{\tau})]}{df} \right|^2 \right\} = 4\pi^2 \sigma_\tau^2 \mathrm{E}\left\{ |H(f)|^2 \right\} \qquad (1.3)$$

where $\mathrm{E}\{\}$ denotes ensemble averaging. In other words, as the RMS delay
spread increases, the channel transfer function fluctuates more wildly over a
particular observation bandwidth in the frequency domain. The basic relationship of Eqn (1.3) is the most important aspect of frequency-selective channels.
Eqn (1.3) is the reason why RMS delay spread is the crucial criterion used
for designing wideband wireless receivers since an increase in delay spread
physically results in more bit errors during demodulation.

While the concepts of frequency selectivity are well understood, they serve
as a useful starting point for the discussion of *spatial selectivity*.

2.3   SPATIAL SELECTIVITY

For a narrowband receiver operating in a static channel, the effects of *spatial selectivity* often limit the performance of a wireless link. The analysis for
characterizing spatial selectivity is nearly identical to that used for frequency
selectivity in the previous section. We characterize the received complex voltage, $\tilde{h}(r)$, as a function of position, r. If the function $\tilde{h}(r)$ is a wide-sense

Table 1.1   Parallel mathematical relationships between spatial and frequency selectivity.

SPACE	FREQUENCY
position, r	frequency, f
wavenumber, k	delay, τ
wavenumber spread, σ_k	delay spread, σ_τ
wavenumber spectrum, $S(k)$	delay spectrum, $p(\tau)$

stationary stochastic process, then it is possible to express its mean-squared rate-of-change as

$$\mathrm{E}\left\{\left|\frac{d[\tilde{h}(r)\exp(-j\overline{k}r)]}{dr}\right|^2\right\} = \sigma_k^2 \mathrm{E}\left\{|\tilde{h}(r)|^2\right\} \tag{1.4}$$

where σ_k is the *wavenumber spread* as given by

$$\sigma_k^2 = \overline{k^2} - (\overline{k})^2, \quad \text{where } \overline{k^n} = \frac{\int\limits_{-\infty}^{+\infty} k^n S(k)dk}{\int\limits_{-\infty}^{+\infty} S(k)dk} \tag{1.5}$$

The function $S(k)$ is the wavenumber spectrum of the space-varying channel, $\tilde{h}(r)$. The left-hand side of Eqn (1.4) is the mean-squared rate-of-change of the space-varying channel, often referred to as the *fading rate variance* [3]. Basically, Eqn (1.4) and Eqn (1.5) state that as the wavenumber spectrum widens, the received signal level fluctuates more wildly in space.

The analysis of spatial selectivity clearly parallels frequency selectivity. Table 1.1 illustrates the parallel mathematical relationships between spatial and frequency selectivity. There are several aspects of spatial selectivity, however, that make analysis more difficult than frequency selectivity. One key difficulty is that the wavenumber spectrum is a function of multipath angle-of-arrival. Thus, the idea of a wavenumber spread must now be cast in terms of angle-of-arrival properties. Furthermore, wavenumber spread is a function of orientation; the same local area may have radically different values of wavenumber spread if the azimuthal orientation of the measurement is changed.

It should be noted that when discussing a mobile receiver with velocity, v, spatial selectivity becomes *temporal* selectivity. Eqn (1.4) and Eqn (1.5) are still valid after the following substitutions are made: 1) change the position dependence, r, to a time dependence, t, by substituting $r = vt$ and 2) change the wavenumber dependence, k, to a Doppler frequency dependence, ω, by

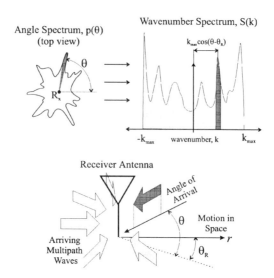

Figure 1.1   Multipath power is mapped from the angle spectrum, $p(\theta)$, to the wavenumber spectrum, $S(k)$, as a function of its angle-of-arrival.

substituting $k = \frac{\omega}{v}$. In the terminology of time-varying channels, wavenumber spectrum becomes *Doppler spectrum* and wavenumber spread becomes *Doppler spread*.

2.4   MAPPING ANGLES TO WAVENUMBERS

The concept of a delay spectrum is intuitive: multipath power arrives with different propagation delays, causing a smear of power as a function of time delay. The concept of a wavenumber spectrum is not as intuitive. Rather, most engineers characterize multipath in space using an *angle* spectrum rather than a wavenumber spectrum. An angle spectrum, $p(\theta)$, describes received power as a function of azimuthal angle-of-arrival, θ.

A simple formula exists for converting an angle spectrum, $p(\theta)$, to a wavenumber spectrum, $S(k)$. The mapping relationship is given by [4] as

$$S(k) = \frac{p(\theta_R + \cos^{-1}\frac{k}{k_{max}}) + p(\theta_R - \cos^{-1}\frac{k}{k_{max}})}{\sqrt{k_{max}^2 - k^2}}, \quad |k| \leq k_{max} \qquad (1.6)$$

where θ_R is the azimuthal direction of movement and k_{max} is the wavenumber of the carrier frequency ($k_{max} = \frac{2\pi}{\lambda}$, where λ is the wavelength of radiation).

The mapping of Eqn (1.6) derives from the geometry of propagation shown in Figure 1.1. A multipath wave arrives from the horizon at angle θ and the direction of azimuthal motion that we wish to map is θ_R. The phase progression of this multipath wave is the free space wavenumber, k_{max}. However, to a receiver moving along the θ_R direction, the actual wavenumber, k, appears to be foreshortened by a factor $\cos(\theta - \theta_R)$. Thus,

$$k = k_{max} \cos(\theta - \theta_R) \quad \text{and} \quad \frac{dk}{d\theta} = -k_{max} \sin(\theta - \theta_R) \qquad (1.7)$$

We arrive at Eqn (1.6) by equating $S(k)|dk| = p(\theta)|d\theta|$. The mapping of Eqn (1.6) provides a useful bridge between spatial selectivity and angle-of-arrival.

Example. In a cluttered multipath environment it is common to approximate the angle spectrum of incoming multipath power as a uniform distribution:

$$p(\theta) = \frac{P_T}{2\pi} \qquad (1.8)$$

where P_T is a constant. Using Eqn (1.6), the wavenumber spectrum for this propagation scenario is

$$S(k) = \frac{P_T}{\pi \sqrt{k_{max}^2 - k^2}} \qquad (1.9)$$

and the wavenumber spread is

$$\sigma_k^2 = \frac{k_{max}^2 P_T}{2} \qquad (1.10)$$

We can now plug this value of wavenumber spread, σ_k, into Eqn (1.4) to gauge the fading rate variance of the space-varying received voltage.

This example serves to illustrate the classical procedure for studying spatial selectivity in a local area:

1. Choose an orientation in space to study, θ_R.

2. Map the angle spectrum, $p(\theta)$, to wavenumber spectrum, $S(k)$.

3. Calculate the wavenumber spread, σ_k.

4. Relate wavenumber spread to the mean-squared spatial fluctuations in received voltage (fading rate variance).

The remaining discussion of this chapter will demonstrate how to relate the geometrical properties directly to the fading rate variance.

3 MULTIPATH SHAPE FACTORS

This section introduces the concept of multipath shape factors – parameters that describe multipath angle-of-arrival characteristics and also imply spatially selective behavior in a multipath channel [5].

3.1   SHAPE FACTOR DEFINITIONS

This section presents the three multipath shape factors that characterize small-scale fading statistics in space. The shape factors are derived from the angular distribution of multipath power, $p(\theta)$, which is a general representation of from-the-horizon propagation in a local area. This representation of $p(\theta)$ includes antenna gains and polarization mismatch effects [6]. Shape factors are based on the complex Fourier coefficients of $p(\theta)$:

$$F_n = \int_0^{2\pi} p(\theta) \exp(jn\theta)d\theta \qquad (1.11)$$

where F_n is the nth complex Fourier coefficient.

3.1.1   Angular Spread.   The shape factor *angular spread*, Λ, is a measure of how multipath concentrates about a single azimuthal direction. We define angular spread to be

$$\Lambda = \sqrt{1 - \frac{|F_1|^2}{F_0^2}} \qquad (1.12)$$

where F_0 and F_1 are defined by Eqn (1.11). There are several advantages to defining angular spread in this manner. First, since angular spread is normalized by F_0 (the total amount of local average received power), it is invariant under changes in transmitted power. Second, Λ is invariant under any series of rotational or reflective transformations of $p(\theta)$. Finally, this definition is intuitive; angular spread ranges from 0 to 1, with 0 denoting the extreme case of a single multipath component from a single direction and 1 denoting no clear bias in the angular distribution of received power.

It should be noted that other definitions exist in the literature for angular spread. These definitions involve either beamwidth or the second centered moment of θ and are often ill-suited for general application to periodic functions such as $p(\theta)$.

3.1.2   Angular Constriction .   The shape factor *angular constriction*, γ, is a measure of how multipath concentrates about *two* azimuthal directions. We define angular constriction to be

$$\gamma = \frac{|F_0 F_2 - F_1^2|}{F_0^2 - |F_1|^2} \qquad (1.13)$$

where F_0, F_1, and F_2 are defined by Eqn (1.11). Much like the definition of angular spread, the measure for angular constriction is invariant under changes in transmitted power or any series of rotational or reflective transformations of $p(\theta)$. The possible values of angular constriction, γ, range from 0 to 1, with 0 denoting no clear bias in two arrival directions and 1 denoting the extreme case of exactly two multipath components arriving from different directions.

3.1.3 Azimuthal Direction of Maximum Fading. A third shape factor, which may be thought of as an orientation parameter, is the *azimuthal direction of maximum fading*, θ_{max}. We define this parameter to be

$$\theta_{max} = \frac{1}{2}\arg\left\{F_0 F_2 - F_1^2\right\} \tag{1.14}$$

The physical meaning of the parameter is presented in the next section.

3.2 BASIC WAVENUMBER SPREAD RELATIONSHIP

Shape factors have a particularly useful application when describing small-scale fading: the multipath angle-of-arrival dependence of wavenumber spread may be cast exclusively in terms of the three shape factors angular spread, angular constriction, and direction of maximum fading. It has been shown that the wavenumber spread for the complex voltage of a receiver traveling along the azimuthal direction θ_R is

$$\sigma_k^2 = \frac{2\pi^2\Lambda^2}{\lambda^2}\left(1 + \gamma\cos\left[2(\theta_R - \theta_{max})\right]\right) \tag{1.15}$$

where λ is the wavelength of the carrier frequency [3]. The value σ_k^2 describes the spatial selectivity of a channel in a local area for a receiver moving in the θ_R direction. Eqn (1.15) is valid for any channel in which multipath waves arrive at the receiver from the horizon – a common assumption when describing mobile radio propagation. The next section discusses the unique aspects of small-scale fading behavior described by each shape factor.

3.3 COMPARISON TO OMNIDIRECTIONAL PROPAGATION

Applying the three shape factors, Λ, γ, and θ_{max}, to the classical omnidirectional propagation model, we find that there is not a bias in either one or two directions of angle-of-arrival, leading to maximum angular spread ($\Lambda = 1$) and minimum angular constriction ($\gamma = 0$). The statistics of omnidirectional propagation are *isotropic*, exhibiting no dependence on the azimuthal direction of receiver travel, θ_R.

If the rate variance relationship of Eqn (1.15) is normalized against their values for omnidirectional propagation, then they reduce to the following form:

$$\sigma^2(\theta_R) = \frac{\sigma_k^2(\theta_R)}{\sigma_{k_{omni}}^2}$$

$$= \Lambda^2 \left(1 + \gamma \cos \left[2(\theta_R - \theta_{max})\right]\right) \qquad (1.16)$$

where σ^2 is a normalized wavenumber spread. Eqn (1.16) provides a convenient way to analyze the effects of the shape factors on the second-order statistics of small-scale fading.

First, notice that angular spread, Λ, describes the *average* fading rate within a local area. A convenient way of viewing this effect is to consider the fading rate variance taken along two perpendicular directions within the same local area. From Eqn (1.16), the average of the two fading rate variances, regardless of the orientation of the measurement, is always given by

$$\frac{1}{2} \left[\sigma^2(\theta_R) + \sigma^2(\theta_R + \pi/2)\right] = \Lambda^2 \qquad (1.17)$$

Eqn (1.17) clearly shows that the average fading rate within a local area decreases with respect to omnidirectional propagation as multipath power becomes more and more concentrated about a single azimuthal direction.

Second, notice that angular constriction, γ, does not affect the average fading rate within a local area, but describes the variability of fading rates taken along different azimuthal directions, θ_R. From Eqn (1.16), fading rate variance σ^2 will change as a function of θ_R, but will always fall within the following range:

$$\sqrt{1 - \gamma} \leq \frac{\sigma(\theta_R)}{\Lambda} \leq \sqrt{1 + \gamma} \qquad (1.18)$$

The upper limit of Eqn (1.18) corresponds to a receiver traveling in the azimuthal direction of maximum fading ($\theta_R = \theta_{max}$) while the lower limit corresponds to travel in a perpendicular direction ($\theta_R = \theta_{max} + \pi/2$). Eqn (1.18) clearly shows that the variability of fading rates within the same local area increases as the channel becomes more and more constricted.

It is interesting to note that the propagation mechanisms of a channel are not uniquely described by the three shape factors Λ, γ, and θ_{max}. An infinitum of propagation mechanisms exist which may have the same set of shape factors and, by extension, lead to channels which exhibit nearly the same end-to-end performance. In fact, Eqn (1.16) provides rigorous mathematical criteria for a multipath channel that may be treated as "pseudo-omnidirectional":

$$|F_1|, |F_2| \ll F_0 \qquad (1.19)$$

Under the condition of Eqn (1.19), angular spread becomes approximately 1 and angular constriction becomes approximately 0. Thus, the second-order statistics of the channel behave nearly identical to the classical omnidirectional channel.

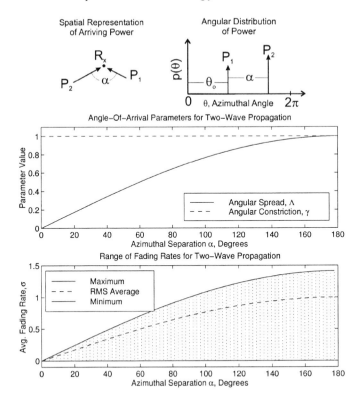

Figure 1.2   Two-wave propagation model. Graphs are for the special case of $P_1 = P_2$.

4   EXAMPLES

This section presents three different analytical examples of directional propagation channels that provide insight into the shape factor definitions and how they describe fading rates. Each example is accompanied by a graph of the angular spread and angular constriction and a graph showing the average and limiting cases of fading rate variance with respect to omnidirectional propagation.

4.1   TWO-WAVE CHANNEL MODEL

Consider the simplest small-scale fading situation where two constant amplitude multipath components, with individual powers defined by P_1 and P_2, arrive at a mobile receiver separated by an azimuthal angle α. Figure 1.2 illustrates this angular distribution of power, which is mathematically defined as

$$p(\theta) = P_1\,\delta(\theta - \theta_o) + P_2\,\delta(\theta - \theta_o - \alpha) \tag{1.20}$$

where θ_o is an arbitrary offset angle and $\delta(\cdot)$ is an impulse function. By applying Eqn (1.12)-Eqn (1.14), the expressions for Λ, γ, and θ_{max} for this distribution are

$$\Lambda = \frac{2\sqrt{P_1 P_2}}{P_1 + P_2} \sin \frac{\alpha}{2}, \qquad \gamma = 1, \qquad \theta_{max} = \theta_0 + \frac{\alpha + \pi}{2} \qquad (1.21)$$

The angular constriction, γ, is always 1 because the two-wave model represents perfect clustering about two directions. The limiting case of two multipath components arriving from the same direction ($\alpha = 0$) results in an angular spread, Λ, of 0. An angular spread of 1 results only when two multipath of identical powers ($P_1 = P_2$) are separated by $\alpha = 180°$. Figure 1.2 shows how the fading behavior changes as multipath separation angle, α, increases for the case of two equal-powered waves. Thus, increasing α changes a channel with low spatial selectivity into a channel with high spatial selectivity that exhibits a strong dependence on the azimuthal direction of receiver motion.

4.2   SECTOR CHANNEL MODEL

Consider another theoretical situation where multipath power is arriving continuously and uniformly over a range of azimuth angles. This model has been used to describe propagation for directional receiver antennas with a distinct azimuthal beam [4]. The function $p(\theta)$ will be defined by

$$p(\theta) = \begin{cases} \frac{P_T}{\alpha} & : \quad \theta_o \leq \theta \leq \theta_o + \alpha \\ 0 & : \quad \text{elsewhere} \end{cases} \qquad (1.22)$$

The angle α indicates the width of the sector (in radians) of arriving multipath power and the angle θ_o is an arbitrary offset angle, as illustrated by Figure 1.3. By applying Eqn (1.12)-Eqn (1.14), the expressions for Λ, γ, and θ_{max} for this distribution are

$$\Lambda = \sqrt{1 - \frac{4 \sin^2 \frac{\alpha}{2}}{\alpha^2}}, \qquad \gamma = \frac{4 \sin^2 \frac{\alpha}{2} - \alpha \sin \alpha}{\alpha^2 - 4 \sin^2 \frac{\alpha}{2}}, \qquad \theta_{max} = \theta_0 + \frac{\alpha + \pi}{2} \qquad (1.23)$$

The limiting cases of these parameters and Eqn (1.15) provide deeper understanding of angular spread and constriction.

Figure 1.3 graphs the spatial channel parameters, Λ and γ, as a function of sector width, α. The limiting case of a single multipath arriving from precisely one direction corresponds to $\alpha = 0$, which results in the minimum angular spread of $\Lambda = 0$. The other limiting case of uniform illumination in all directions corresponds to $\alpha = 360°$ (omnidirectional Clarke model), which results in the maximum angular spread of $\Lambda = 1$. The angular constriction, γ, follows an opposite trend. It is at a maximum ($\gamma = 1$) when $\alpha = 0$ and at a

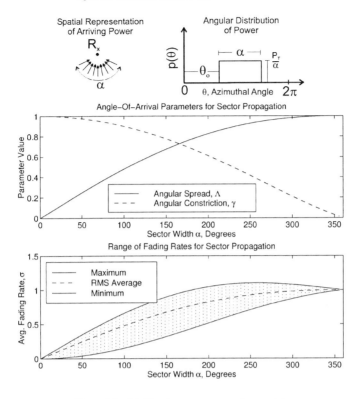

Figure 1.3 Multipath sector propagation model.

minimum ($\gamma = 0$) when $\alpha = 360°$. The graph in Figure 1.3 shows that as the multipath angles of arrival are condensed into a smaller and smaller sector, the directional dependence of fading rates within the same local area increases. Overall, however, fading rates decrease with decreasing sector size α.

4.3 DOUBLE SECTOR CHANNEL MODEL

Another example of angular constriction may be studied using the Double Sector model of Figure 1.4. Diffuse multipath propagation over two equal and opposite sectors of azimuthal angles characterize the incoming power. The equation that describes this angular distribution of power is

$$p(\theta) = \begin{cases} \frac{P_T}{2\alpha} & : \quad \theta_o \leq \theta \leq \theta_o + \alpha, \quad \theta_o + \pi \leq \theta \leq \theta_o + \alpha + \pi \\ 0 & : \quad \text{elsewhere} \end{cases} \quad (1.24)$$

The angle α is the sector width and the angle θ_o is an arbitrary offset angle. By applying Eqn (1.12)-Eqn (1.14), the expressions for Λ, γ, and $\theta_{\max}$ for this

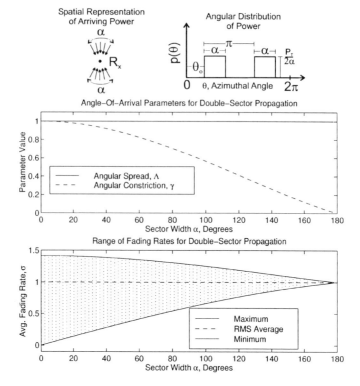

Figure 1.4   Multipath double sector propagation model.

distribution are

$$\Lambda = 1, \qquad \gamma = \frac{\sin \alpha}{\alpha}, \qquad \theta_{max} = \theta_0 + \frac{\alpha}{2} \qquad (1.25)$$

Note that the value of angular spread, Λ, is always 1. Regardless of the value of α, an equal amount of power arrives from opposite directions, producing no clear bias in the direction of multipath arrival.

The limiting case of $\alpha = 180°$ (omnidirectional propagation) results in an angular constriction of $\gamma = 0$. As α decreases, the angular distribution of power becomes more and more constricted. In the limit of $\alpha = 0$, the value of angular constriction reaches its maximum, $\gamma = 1$. This case corresponds to the above-mentioned instance of two-wave propagation. Figure 1.4 shows how the fading behavior changes as sector width α increases, making the fading rate more and more isotropic while the RMS average remains constant.

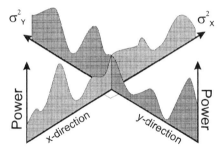

Figure 1.5   Power measured along two perpendicular linear tracks in space within a local area.

5   APPLICATIONS

This section presents three examples of how multipath shape factors may be used to solve practical problems in channel modeling and measurement.

5.1   MEASUREMENT OF ANGULAR SPREAD

Problem Statement.   A researcher is equipped with a simple non-coherent receiver connected to an omnidirectional antenna. Since the receiver is non-coherent, it is only capable of measuring received power (no phase). How can angular spread be measured with such a simple receiver configuration?

Solution.   A procedure for measuring angular spread, Λ, may be based on measuring received power along two perpendicular directions in space within a local area. Much like Eqn (1.17), the mean-squared rate-of-change of power measured along two perpendicular directions may be summed to a result proportional to angular spread [6]. Thus, if σ_x^2 and σ_y^2 represent the mean-squared rates of power change measured along the x and y directions, respectively, then angular spread is given by

$$\Lambda^2 = \frac{\lambda^2}{8\pi^2 P_T^2} \left[\sigma_x^2 + \sigma_y^2 \right] \tag{1.26}$$

where λ is radiation wavelength and P_T is the spatially-averaged power level in the local area. Figure 1.5 illustrates this approach. The result in Eqn (1.26) is independent of the measurement orientation in azimuth: the only requirement is that the two tracks be perpendicular. This approach was first used by [7] to characterize wideband small-scale fading for outdoor transmitter-receiver configurations at 1900 MHz.

5.2   LEVEL-CROSSING STATISTICS

Problem Statement.   A mobile wireless receiver experiences small-scale fading with temporal statistics that depend on wavelength, receiver velocity, and multipath angle-of-arrival. The level-crossing rate, N_R, of a fading channel is the average rate that the received voltage envelope crosses a specified threshold level. Similarly, the average fade duration, $\bar{\tau}$, is the average time interval that received voltage envelope spends below the threshold level each time that level is crossed.

Both level-crossing rate and average fade duration are used to describe how often an acceptable *signal-to-noise ratio* is maintained in a fading channel. Thus, both are critical parameters that affect the capacity and performance of the fading wireless link. If a mobile receiver travels with speed, v, in the azimuthal direction, θ_R, and experiences small-scale Rayleigh fading, then what are the general expressions for level-crossing rate and average fade duration?

Solution.   The solutions for level-crossing rate and average fade duration for an *omnidirectional* multipath channel is described by Jakes [8]. Recently, it has been shown that the most general solution for level-crossing rate and average fade duration may be expressed for any arbitrary from-the-horizon multipath channel in terms of the three basic shape factors [9]. For a threshold level ρ, where ρ is the ratio of voltage threshold to the RMS received voltage in the channel, the level-crossing rate is given by

$$N_R = \frac{\sqrt{2\pi}\,v\Lambda\rho}{\lambda}\sqrt{1 + \gamma\cos\left[2(\theta_R - \theta_{max})\right]}\exp\left(-\rho^2\right) \qquad (1.27)$$

where λ is the carrier wavelength. The average fade duration for the same threshold, ρ, is

$$\bar{\tau} = \frac{\lambda\left[\exp\left(\rho^2\right) - 1\right]}{\sqrt{2\pi}\,v\rho\Lambda\sqrt{1 + \gamma\cos\left[2(\theta_R - \theta_{max})\right]}} \qquad (1.28)$$

For Rayleigh fading channels, both Eqn (1.27) and Eqn (1.28) are exact. These equations may be used to study the effects of non-omnidirectional multipath and directional antennas on mobile fading statistics.

5.3   ENVELOPE DECORRELATION BETWEEN ANTENNA ELEMENTS

Problem Statement.   A narrowband receiver with multiple antennas may employ space diversity techniques to combat small-scale fading. Maximum diversity gain, however, is only achieved if the fading on each antenna element is uncorrelated with the others [10]. If two omnidirectional antennas are used at the receiver for diversity, what is the optimal orientation and spacing between

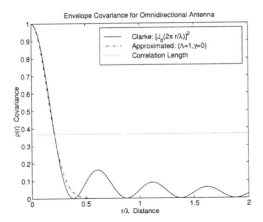

Figure 1.6 Comparison between Clarke theoretical and approximate envelope autocovariance functions for omnidirectional propagation.

the antennas in a Rayleigh flat-fading environment that allows for uncorrelated fading?

Solution. It has been shown that, although spatial correlation functions for voltage envelope vary significantly from case-to-case, the behavior of these functions over small distances is described accurately in terms of shape factors [3]. A useful approximation to the envelope autocovariance function in a Rayleigh flat-fading channel is given by

$$\rho(r, \theta_R) \approx \exp\left[-23\Lambda^2 \left(1 + \gamma \cos[2(\theta_R - \theta_{\max})]\right)\left(\frac{r}{\lambda}\right)^2\right] \qquad (1.29)$$

where r is the separation distance between two points in space, θ_R is their azimuthal orientation, and λ is the wavelength of radiation. Eqn (1.29) captures the basic correlation behavior between antenna elements in space, asymptotically decreasing towards 0 for increasing values of r. An example demonstrating the accuracy of this approximation may be seen in Figure 1.6. In this comparison, Eqn (1.29) for omnidirectional propagation ($\Lambda = 1$, $\gamma = 0$) is plotted against a known analytical solution [8]. Note that, although the higher-order behavior is not modeled by Eqn (1.29), the approximation is very accurate for small values of separation distance, r.

The approximate behavior of Eqn (1.29) is accurate enough to estimate the correlation length of the voltage envelope. A correlation length, D_c, is the distance which satisfies the relationship $\rho(D_c, \theta_R) = \exp(-1)$ for a given orientation in azimuth, θ_R. Fading envelopes with correlation less than $\exp(-1)$ are essentially uncorrelated for the purposes of space diversity [11]. From

Eqn (1.29), it is possible to express the correlation criterion in terms of shape factors:

$$D_c \approx \frac{\lambda}{\Lambda\sqrt{23(1 + \gamma \cos[2(\theta_R - \theta_{max})])}} \qquad (1.30)$$

We now can derive the design criterion for 2-element space diversity from Eqn (1.30). First, we see that to achieve the smallest separation between the two antennas, they should be oriented along the azimuthal direction of maximum fading, $\theta_R = \theta_{max}$. Eqn (1.30) then reduces to

$$D_c \approx \frac{\lambda}{\Lambda\sqrt{23(1 + \gamma)}} \qquad (1.31)$$

which provides the minimum separation distance for the 2 diversity antennas under the optimum orientation. Only knowledge of angular spread, Λ, and angular constriction, γ, must be known to calculate the correlation length.

6   SUMMARY

This chapter has presented a theoretical framework for relating multipath angle-of-arrival characteristics to the spatially-selective behavior of small-scale fading. The framework characterizes the multipath angle-of-arrival using geometrical shape factors. The theory of multipath shape factors may be summarized by the following key points:

1. Spatial selectivity in a received signal is caused by multipath waves arriving from different directions in space, similar to the frequency selectivity caused by multipath waves arriving with different time delays.

2. The dominant behavior of spatial selectivity may be captured by three *shape factors*: angular spread Λ, angular constriction γ, and azimuthal direction of maximum fading θ_{max}.

3. The shape factors represent simple geometrical properties of the multipath angles-of-arrival in a local area.

4. Each shape factor also represents a type of small-scale fading behavior exhibited by the channel.

5. Most practical channel modeling problems that involve spatial selectivity may be solved in terms of the shape factors of multipath propagation.

Several useful applications of the shape factor theory were presented. There are numerous other applications for multipath shape factors in wireless communications, although most fall within the following three areas:

- **Channel Measurement:** Shape factors provide a great deal of insight on how to design radio frequency measurement campaigns that measure small-scale fading and multipath angle-of-arrival. The theory demonstrates how spatial selectivity may be implied from angle-of-arrival measurements and vice versa.

- **Channel Modeling:** Basic fading statistics (fading rate variance, level-crossing rate, average fade duration, etc.) may be expressed in terms of the multipath shape factors. The three shape factors emphasize what is important about a channel model (the gross shape of the incoming multipath, low-order Fourier coefficients of the angle spectrum) and what information may be ignored or left unmodeled (the fine structure of incoming multipath, high-order Fourier coefficients of the angle spectrum). The geometrically intuitive shape factors allow for quick insight into how different propagation scenarios or receiver antenna patterns affect small-scale fading.

- **Design Criterion:** The design criterion for systems affected by spatial selectivity may often be expressed in terms of angular spread, angular constriction, and the direction of maximum fading. For example, if a space-time coding algorithm requires uncorrelated fading at the receiver antennas, then a criterion similar to Eqn (1.31) could be used to place the antenna elements.

In summary, the multipath shape factors provide a theoretical medium for analysis and exploration of new spatial channel models for wireless systems of the future.

References

[1] G. Foschini, "Layered Space-Time Architecture for Wireless Communication in a Fading Environment When Using Multi-Element Antennas," *Bell Labs Technical Journal*, pp. 41–59, Autumn 1996.

[2] T. Rappaport, *Wireless Communications: Principles and Practice*. New Jersey: Prentice-Hall Inc., 1996.

[3] G. Durgin and T. Rappaport, "Theory of Multipath Shape Factors for Small-Scale Fading Wireless Channels," *to appear in IEEE Transactions on Antennas and Propagation*, Apr 2000.

[4] M. Gans, "A Power-Spectral Theory of Propagation in the Mobile Radio Environment," *IEEE Transactions on Vehicular Technology*, vol. VT-21, pp. 27–38, Feb 1972.

[5] G. Durgin and T. Rappaport, "Three Parameters for Relating Small-Scale Temporal Fading to Multipath Angles-of-Arrival," in *PIMRC '99*, (Osaka Japan), pp. 1077–1081, Sep 1999.

[6] G. Durgin and T. Rappaport, "A Basic Relationship Between Multipath Angular Spread and Narrowband Fading in a Wireless Channel," *IEE Electronics Letters*, vol. 34, pp. 2431–2432, 10 Dec 1998.

[7] N. Patwari, G. Durgin, T. Rappaport, and R. Boyle, "Peer-to-Peer Low Antenna Outdoor Radio Wave Propagation at 1.8 GHz," in *49th IEEE Vehicular Technology Conference*, vol. 1, (Houston TX), pp. 371–375, May 1999.

[8] W. Jakes, *Microwave Mobile Communications*. New York: IEEE Press, 1974.

[9] G. Durgin and T. Rappaport, "Level Crossing Rates and Average Fade Duration of Wireless Channels with Spatially Complicated Multipath," in *Globecom '99*, (Brazil), Dec 1999.

[10] W. Jakes, "A Comparison of Specific Space Diversity Techniques for Reduction of Fast Fading in UHF Mobile Radio Systems," *IEEE Transactions on Vehicular Technology*, vol. VT-20, pp. 81–91, Nov 1971.

[11] D. Reudink, "Properties of Mobile Radio Propagation Above 400 MHz," *IEEE Transactions on Vehicular Technology*, Nov 1974.

Chapter 2

SPACE-TIME CODING AND SIGNAL PROCESSING FOR HIGH DATA RATE WIRELESS COMMUNICATIONS

Ayman. F. Naguib
AT& T Labs - Research 180 Park Avenue Florham Park, NJ 07932
naguib@research.att.com

Rob Calderbank
AT& T Labs - Research 180 Park Avenue Florham Park, NJ 07932
rc@research.att.com

Abstract   The information capacity of wireless communication systems can be increased dramatically by employing multiple transmit and receive antennas [?, ?]. An effective approach to increasing data rate over wireless channels is to employ coding techniques appropriate to multiple transmit antennas, that is space-time coding. Space-time codes introduce temporal and spatial correlation into signals transmitted from different antennas, in order to provide diversity at the receiver, and coding gain over an uncoded system. The spatial-temporal structure of these codes can be exploited to further increase the capacity of wireless systems with a relatively simple receiver structure. This chapter provides an overview of space-time coding techniques and the associated signal processing framework.

Keywords: Space-Time Coding, Array Signal Processing, Interference Suppression, OFDM, Wireless Communications

1   INTRODUCTION

The goal of high data rate wireless communication between two portable terminals that may be located anywhere in the world, and the vision of a single phone that acts as a traditional cellular phone when used outdoors, and as a conventional high quality phone when used indoors [3], are driving recent developments in communications. The great popularity of cordless phones, cellular phones, radio paging, portable computing, and other personal communication services (PCS) demonstrates rising demand for these services. Rapid growth in mobile computing and other wireless data services is inspiring many proposals for high speed data services in the range of 64-144 kbps for micro cellular wide area and high mobility applications, and up to 2 Mbps for indoor applications [4]. In addition to mobile applications, fixed wireless access (FWA) technologies offer the promise of bringing high quality telephony, high speed internet access, multi-media, and other broadband services to the home over wireless links [5, 6]. Research challenges in this area include the development of efficient coding and modulation, signal processing techniques to improve the quality and spectral efficiency of wireless communications, and better techniques for sharing the limited spectrum among different high capacity users.

The physical limitation of the wireless channel presents a fundamental technical challenge for reliable communications. The channel is subject to time-varying impairments such as noise, interference, and multipath [7, 8, 9, 10, 11, 12, 13]. Limitations on the power and size of the mobile terminal and of network terminating devices (NTD) in a FWA application is a second major design consideration. Most personal communications and wireless services portables are meant to be carried in a briefcase and/or pocket and must, therefore, be small and lightweight, which translates to a low power requirement since small batteries must be used. Although a NTD in FWA applications may have more signal processing power than a mobile computing portable, power consumption and device and antenna size are still a concern. However, many of the signal processing techniques which may be used for reliable communications and efficient spectral utilization demand significant processing power, precluding the use of low power devices. Continuing advances in VLSI and application-specific integrated circuit (ASIC) technology for low power applications will provide a partial solution to this problem. Hence, placing more signal processing burden on fixed locations (base stations) with relatively larger power resources than the portable makes good engineering sense.

2   DIVERSITY TECHNIQUES

Several diversity techniques have been employed in wireless communication systems to improve the link margin. Diversity techniques which may be used include time, frequency, and space diversity

- *Time diversity*: channel coding in combination with limited interleaving is used to provide time diversity. However, while channel coding is extremely effective in fast fading environments (high mobility), it offers very little protection under slow fading (low mobility and FWA) unless significant interleaving delays can be tolerated.

- *Frequency diversity*: the fact that signals transmitted over different frequencies induce different multipath structure and independent fading is exploited to provide frequency diversity (sometimes referred to as path diversity). In TDMA systems, frequency diversity is obtained by the use of equalizers [14] when the multipath delay spread is a significant fraction of a symbol period. GSM uses frequency hopping to provide frequency diversity. In DS-CDMA systems, RAKE receivers [15, 16] are used to obtain path diversity. However, when the multipath delay spread is small, compared to the symbol period, frequency or path diversity does not exist.

- *Space diversity*: the receiver/transmitter uses multiple antennas that are separated and/or differently polarized for reception/transmission to create independent fading channels. Currently, multiple antennas at base-stations are used for receive diversity at the base, However, it is difficult to have more than one or two antennas at the portable unit due to the size limitations and cost of multiple chains of RF down conversion.

Both receive and polarization diversity have received a lot of attention [11, 12, 17]. In fact, in current cellular applications, receive diversity is already used for improving reception from mobiles. In polarization diversity, two antennas with different polarization are used to receive (or transmit) the signal. Different polarization will ensure that the fading channel corresponding to each of the two antennas will be independent without having to place the two antennas far apart. In receive diversity, two or more antennas that are well separated (again, to ensure independent fading channels) are used to generate independent looks at the transmitted signal. These different variants of the transmitted signal can be processed in several ways to improve the overall signal quality. In selection diversity, the best received signal is used, and this signal can be chosen based on several quality metrics, including total received power, signal-to-noise ratio (SNR), etc. Another form of selection diversity is switched diversity in which an alternate antenna is chosen if the received signal level falls below

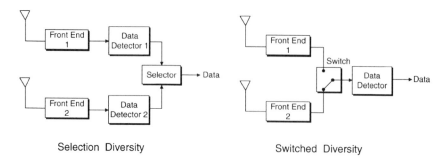

Figure 2.1   Selection and Switched Diversity.

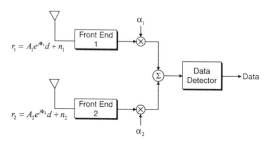

Figure 2.2   Linear Combining Diversity.

a certain threshold. Figure 2.1 shows a block diagram of selection diversity schemes.

The second form of receive diversity is based on linear combining. As the name implies, the signal used for detection in linear combining techniques is a linear combination of a weighted replica of all received signals. Figure 2.2 shows a block diagram for receive diversity with linear combining and two receive antennas. In this block diagram, let r_1 and r_2 be the received signals at antennas 1 and 2, respectively, where

$$r_1 = A_1 e^{j\phi_1} d + n_1 \qquad (2.1)$$
$$r_2 = A_2 e^{j\phi_2} d + n_2 \qquad (2.2)$$

d is the information symbol, n_1 and n_2 are the additive white Gaussian noise at antenna 1 and 2 respectively, and A_i and ϕ_i $(i = 1, 2)$ is the corresponding amplitude and phase of the fading channel, respectively. The receiver uses the linear combination $\tilde{r} = \alpha_1 r_1 + \alpha_2 r_2$. The weighting coefficients α_1, α_2 can be chosen in several ways. In **equal gain combining** the weights are chosen as $\alpha_1 = e^{-j\phi_1}$ and $\alpha_2 = e^{-j\phi_2}$. In this way the two antenna signals are co-

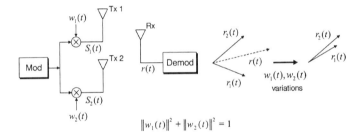

Figure 2.3   Transmit Diversity with Feedback.

phased and added together. A second approach is **maximal ratio combining**, where the two signals are also weighted with their corresponding amplitudes A_1 and A_2. In this case $\alpha_1 = A_1 e^{-j\phi_1}$ and $\alpha_2 = A_2 e^{-j\phi_2}$. A third approach is **minimum mean squared error (MMSE) combining**, where the weighting coefficients are chosen during a training phase such that

$$(\alpha_1, \alpha_2) = \arg \min_{\alpha_1, \alpha_2} |\alpha_1 r_1 + \alpha_2 r_2 - d|^2 \qquad (2.3)$$

The performance of MMSE linear combining and maximal ratio combining are essentially the same. In general, there will be a dramatic improvement in the average SNR, even with 2 branch selection diversity. For all the above approaches to receive diversity, the average SNR will increase with the number of receive antennas. However, for the selection diversity, the SNR increases very slowly with the number of receive antennas. For maximal ratio combining, the average SNR will increase linearly with the number of receive antennas. For equal gain combining the rate of SNR increase will be slightly less than that of maximal ratio combining. In fact, the difference between the two is only 1.05 dB in the limit of an infinite number of receive antennas [18].

Transmit diversity on the other hand has received comparatively little attention. The information theoretic aspects of transmit diversity were addressed in [1, 2, 19, 20]. Previous work on transmit diversity can be classified into three broad categories: schemes using feedback, schemes with feedforward or training information but no feedback, and blind schemes. The first category uses feedback, either explicitly or implicitly, from the receiver to the transmitter to train the transmitter. Figure 2.3 shows a conceptual block diagram for transmit diversity with feedback. A signal is weighted differently and transmitted from two different antennas. The weights w_1 and w_2 are varied such that the received signal power $|r(t)|^2$ is maximized. The weights are adapted based on feedback information from the receiver. For instance, in time division duplex (TDD) systems [21], the same antenna weights are used for reception and transmission, so feedback is implicit in the exploitation of channel sym-

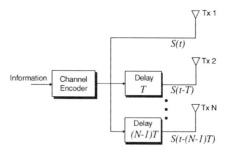

Figure 2.4   Transmit Delay Diversity.

metry. These weights are chosen during reception to maximize the receive signal to noise ratio, and during transmission to weight the amplitudes of the transmitted signals, and, therefore, will also maximize the signal to noise ratio at the receiver. Explicit feedback includes switched diversity systems with feedback [22]. However, in practice, movement by either the transmitter or the receiver (or the surroundings such as cars) and interference dynamics causes a mismatch between the channel perceived by the transmitter and that perceived by the receiver.

Transmit diversity schemes mentioned in the second category use linear processing at the transmitter to spread the information across antennas. At the receiver, information is recovered by an optimal receiver. Feedforward information is required to estimate the channel from the transmitter to the receiver. These estimates are used to compensate for the channel response at the receiver. The first scheme of this type is the delay diversity scheme (see Fig. 2.4) proposed by Wittneben [23] and it includes the delay diversity scheme of [24] as a special case. The linear processing techniques were also studied in [25, 26]. It was shown in [27, 28] that delay diversity schemes are indeed optimal in providing diversity, in the sense that the diversity gain experienced at the receiver (which is assumed to be optimal) is equal to the diversity gain obtained with receive diversity. The delay diversity scheme can be viewed as creating an intentional multipath which can exploited at the receiver by using an equalizer. The linear filtering used (to create delay diversity) at the transmitter can also be viewed as a channel code that takes binary or integer input and creates real valued output. The advantage of delay diversity over other transmit diversity schemes is that it will achieve the maximum possible diversity order (i.e. number of transmit antennas) without any sacrifice in the bandwidth.

The third category does not require feedback or feedforward information. Instead, it uses multiple transmit antennas combined with channel coding to provide diversity. An example of this approach is the use of channel cod-

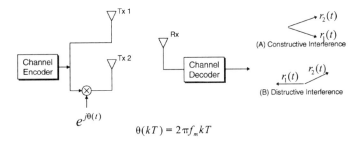

$$\theta(kT) = 2\pi f_m kT$$

Figure 2.5   Transmit Diversity with Phase Sweeping.

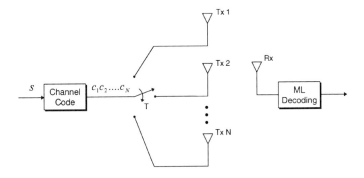

Figure 2.6   Transmit Diversity with Channel Coding.

ing along with phase sweeping [29] or of frequency offset [30] with multiple transmit antennas, to simulate fast fading, as shown in Figure 2.5. An appropriately designed channel code/interleaver pair is used to provide the diversity benefit. Another approach in this category is to encode information by a channel code (Figure 2.6) and to transmit the code symbols using different antennas in an orthogonal manner. This can be done by either time multiplexing [29], or by using orthogonal spreading sequences for different antennas [31]. The disadvantage of these schemes as compared to the previous two categories is the loss in bandwidth efficiency due to the use of the channel code. Using appropriate coding, it is possible to relax the orthogonality requirement needed in these schemes and to obtain the diversity as well as a coding gain without sacrificing bandwidth. This will be possible to do if one views the whole system as a multiple input/multiple output system and uses channel codes that are designed with that view in mind.

In general, all transmit diversity schemes described above can be represented by a single transmitter structure as shown in Figure 2.7. By appro-

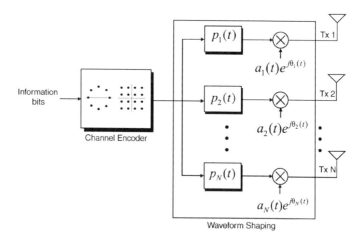

Figure 2.7   General Structure of Transmit Diversity Schmes.

priately selecting the pulse shaping function $p_i(t)$ and the weight $a_i(t)e^{j\theta_i(t)}$ we can obtain any of the above transmit diversity schemes. For example, the delay diversity scheme can be obtained from the above structure by setting all the weights to 1 and the pulse shaping functions to simple time shifts. Note that for, all the transmit diversity schemes in third category, any channel code could be used. As pointed out, the use of a channel code in combination with multiple transmit antennas would achieve diversity, but will suffer a loss in bandwidth (due to channel coding). However by using channel codes that are specifically designed for multiple transmit antennas, one can achieve the needed diversity gain without any sacrifice in bandwidth. These codes are called **Space-Time Codes** (STC). Space-Time coding [32, 33, 34, 35, 36, 37, 38, 39, 40, 41, 42, 43, 44] is a coding technique that is designed for use with multiple transmit antennas. Space-time codes introduce temporal and spatial correlation into signals transmitted from different antennas, so as to provide diversity at the receiver, and coding gain over an uncoded systems without sacrificing the bandwidth. The spatial-temporal structure of these codes can be exploited to further increase the capacity of wireless systems with a relatively simple receiver structure [45]. In the next section we will review space-time coding (STC) and its associated signal processing framework.

3   SPACE-TIME CODING

In this section, we will describe a basic model for a communication system that employs space time coding with N transmit antennas and M receive

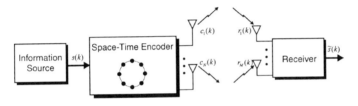

Figure 2.8 Space-Time Coding.

antennas. As shown in Figure 2.8, the information symbol $s(l)$ at time l is encoded by the ST encoder as N code symbols $c_1(l)$, $c_2(l)$, $\cdots$, $c_N(l)$. Each code symbol is transmitted, *simultaneously*, from a different antenna. The encoder chooses the N code symbols to transmit so that both the coding gain and diversity gain at the receiver are maximized.

Signals arriving at different receive antennas undergo independent fading. The signal at each receive antenna is a noisy superposition of the faded versions of the N transmitted signals. A flat fading channel is assumed. We assume that the signal constellation is scaled so that the average energy of the constellation points is 1. Also, Let us assume that E_s is the total energy transmitted (from all antennas) per input symbol. Therefore, the energy per input symbol transmitted from each transmit antenna is E_s/N. Let $r_j(l)$, $j = 1 \cdots M$ be the received signal at antenna j after matched filtering. Assuming ideal timing and frequency information, we have

$$r_j(l) = \sqrt{E_s/N} \cdot \sum_{i=1}^{N} h_{ij}(l)c_i(l) + \eta_j(l), \qquad j = 1, \cdots, M \qquad (2.4)$$

where $\eta_j(l)$ are independent samples of a zero mean complex white Gaussian process with two sided power spectral density $N_0/2$ per dimension. It is also assumed that $\eta_j(l)$ and $\eta_k(l)$ are independent for $j \neq k, 1 \leq j, k \leq M$. The gain $h_{ij}(l)$ models the complex fading channel gain from transmit antenna i to receive antenna j. It is assumed that $h_{ij}(l)$ and $h_{qk}(l)$ are independent for $i \neq q$ or $j \neq k, 1 \leq i, q \leq N, 1 \leq j, k \leq M$. This condition is satisfied if the transmit antennas are *well* separated (by more than $\lambda/2$) or by using antennas with different polarization.

Let $\mathbf{c}_l = [c_1(l), \cdots, c_N(l)]^T$ be the $N \times 1$ *code vector* transmitted from the N antennas at time l, $\boldsymbol{h}_j(l) = [h_{1j}(l), \cdots, h_{Nj}(l)]^T$ be the corresponding $N \times 1$ channel vector from the N transmit antennas to the jth receive antenna, and $\mathbf{r}(l) = [r_1(l), \cdots, r_M(l)]^T$ be the $M \times 1$ received signal vector. Also, let $\boldsymbol{\eta}(l) = [\eta_1(l), \cdots, \eta_M(l)]^T$ be the $M \times 1$ noise vector at the receive antennas. Let us define the $M \times N$ channel matrix $\mathcal{H}_l$ from the N transmit to the M receive antennas as $\mathcal{H}(l) = [\boldsymbol{h}_1(l), \cdots, \boldsymbol{h}_M(l)]^T$. Equation 2.4 can be rewritten in a matrix form as

$$r(l) = \sqrt{E_s/N} \cdot \mathcal{H}(l) \cdot c_l + \eta(l) \, . \tag{2.5}$$

We can easily see that the *signal to noise ratio* (SNR) *per receive antenna* is given by

$$\text{SNR} = \frac{E_s}{N_o} \tag{2.6}$$

4   SPACE-TIME TRELLIS CODES

Suppose that the *code vector* sequence

$$\mathcal{C} = c_1, \, c_2, \cdots, \, c_L$$

was transmitted. We consider the probability that the decoder decides erroneously in favor of the legitimate code vector sequence

$$\tilde{\mathcal{C}} = \tilde{c}_1, \, \tilde{c}_2, \cdots, \, \tilde{c}_L.$$

Consider a frame or block of data of length L and define the $N \times N$ error matrix $\mathcal{A}$ as

$$\mathcal{A}(\mathcal{C}, \tilde{\mathcal{C}}) = \sum_{l=1}^{L} (c_l - \tilde{c}_l)(c_l - \tilde{c}_l)^* \, , \tag{2.7}$$

where $(.)^*$ denotes the conjugate operation for scalers and the conjugate transpose for matrices and vectors. If ideal channel state information (CSI) $\mathcal{H}(l)$, $l = 1, \cdots, L$ is available at the receiver, then it is straightforward to show that the probability of transmitting $\mathcal{C}$ and deciding in favor of $\tilde{\mathcal{C}}$ is upper bounded by [46]

$$P(\mathcal{C} \rightarrow \tilde{\mathcal{C}}) \leq \left(\prod_{i=1}^{r} \lambda_i \right)^{-M} \cdot (E_s/4N_o)^{-rM} \, , \tag{2.8}$$

where r is the rank of the error matrix $\mathcal{A}$ and λ_i, $i = 1, \cdots, r$ are the nonzero eigenvalues of the error matrix $\mathcal{A}$. We can easily see that the probability of error bound in (2.8) is similar to the probability of error bound for trellis coded modulation for fading channels. The first term $g_r = \prod_{i=1}^{r} \lambda_i$ represents the coding gain achieved by the space-time code and the second term $(E_s/4N_o)^{-rM}$ represents a diversity gain of rM. It is clear that in designing a space-time trellis code, the rank of the error matrix r should be maximized (thereby maximizing the diversity gain) and at the same time g_r should be also maximized (thereby maximizing the coding gain).

As an example for space-time trellis codes, we provide an 8-PSK 8-state ST code designed for 2 transmit antennas. Figure 2.9 provides a labeling of the 8-PSK constellation and a the trellis description for this code. Each row

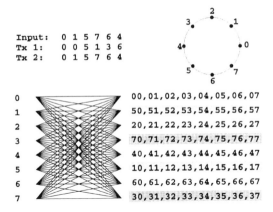

8-PSK 8-State Space-Time Code with 2 Tx Antennas

Figure 2.9 8-PSK 8-state space-time code with 2 transmit antennas.

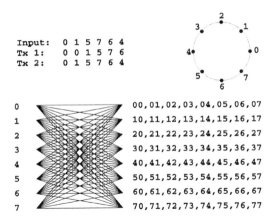

8-PSK 8-State Delay Diversity Code with 2 Tx Antennas

Figure 2.10 8-PSK 8-state delay diversity as a space-time code.

in the matrix shown in Figure 2.9 represents the edge labels for transitions from the corresponding state. The edge label $s_1 s_2$ indicates that symbol s_1 is transmitted over the first antenna and that symbol s_2 is transmitted over the second antenna. The input bit stream to the ST encoder is divided into groups of 3 bits and each group is mapped into one of 8 constellation points. This code has a bandwidth efficiency of 3 bits/channel use.

Figure 2.10 shows the space-time coding representation of delay diversity. It is also interesting to note that the two trellis codes in Figures 2.9 and 2.10 are similar. In fact, we can get the code in Figure 2.9 by swapping the row that starts with a "1" with the row that starts with a "5" and the row that starts with a "3" with the row that starts with a "7". By looking at the constellation points in Figure 2.9, we will easily realize that this space-time code is delay-diversity except that the delayed symbol is multiplied by -1 if it is an odd symbol $\{1, 3, 5, 7\}$ and by +1 if it is an even symbol $\{0, 2, 4, 6\}$. This simple mapping of the delayed symbol gives a 2.5 dB of coding gain as compared to simple delay diversity.

As we mentioned above, delay diversity can be viewed as a space time code and, therefore, the performance analysis presented above applies to it. Consider the delay diversity of [24, 25] where the channel encoder is a rate $1/2$ block repetition code defined over some signal alphabet. Let $\bar{c}_1(l)\bar{c}_2(l)$ be the output of the channel encoder, where $\bar{c}_1(l)$ is to be transmitted from antenna 1, and $\bar{c}_2(l)$ is to be transmitted from antenna 2 one symbol later. This can be viewed as a space-time code by defining the *code vector* $\mathbf{c}(l)$ as

$$\mathbf{c}_l = \left(\begin{array}{c} c_1(l) \\ \bar{c}_2(l-1) \end{array} \right) . \tag{2.9}$$

The minimum determinant of this code is $(2 - \sqrt{2})^2$. Next, consider the block code

$$\mathcal{C} = \{00, 15, 22, 37, 44, 51, 66, 73\} \tag{2.10}$$

of length 2 defined over the 8-PSK alphabet instead of the repetition code. This block code is the best in the sense of product distance [24] amongst all the codes of cardinality 8 and of length 2 defined over the 8-PSK alphabet. This means that the minimum of the product distance $|c_1 - \tilde{c}_1||c_2 - \tilde{c}_2|$ between pairs of distinct codewords $\mathbf{c} = c_1 c_2 \in \mathcal{C}$ and $\tilde{\mathbf{c}} = \tilde{c}_1 \tilde{c}_2 \in \mathcal{C}$ is the maximum amongst all such codes. A delay diversity code constructed from this repetition code is identical to the 8-PSK 8-state space-time code [36]. The minimum determinant of this code is 2.

For decoding space-time codes, we assume that the channel information $\mathcal{H}(l)$, $l = 1, \cdots, L$ is available at the receiver. Suppose that a code vector sequence $\mathcal{C} = \mathbf{c}_1, \mathbf{c}_2, \cdots, \mathbf{c}_L$ has been transmitted, and $\mathcal{R} = \mathbf{r}_1, \mathbf{r}_2, \cdots, \mathbf{r}_L$ has been received, where $\mathbf{r}_l$ is given by (2.5). At the receiver, optimum decoding amounts to choosing a vector code sequence $\tilde{\mathcal{C}} = \tilde{\mathbf{c}}_1, \tilde{\mathbf{c}}_2, \cdots, \tilde{\mathbf{c}}_L$ for which the *a posteriori* probability

$$\Pr\left(\tilde{\mathcal{C}}|\mathcal{R}, \mathcal{H}(l), l = 1, \cdots, L\right)$$

is maximized. Assuming that all codewords are equiprobable, then since the noise vector is assumed to be a multivariate AWGN, it can be easily shown that

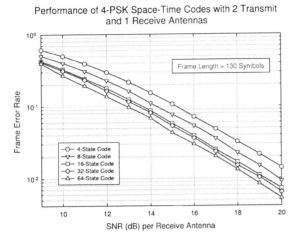

Figure 2.11 Performance of 4-PSK Space-Time Trellis Codes with 2 Transmit and 1 Receive Antennas.

the optimum decoder is [46]

$$\tilde{\mathcal{C}} = \arg \min_{\mathcal{C} = \tilde{c}_1, \cdots, \tilde{c}_L} \sum_{l=1}^{L} \left\| \mathbf{r}(l) - \sqrt{E_s} \cdot \mathcal{H}(l) \cdot \tilde{c}_l \right\|^2 . \qquad (2.11)$$

For the space-time codes with trellis representations (as in the example in Figure 2.9), it is obvious that the optimum decoder in (2.11) can be implemented using the Viterbi algorithm. Note that knowledge of the channel is required for decoding. The receiver, therefore, must estimate the channel either blindly or by using pilot/training symbols. Figure 2.11 shows the performance of 4-PSK space-time trellis codes for 2 transmit and 1 receive antennas with different numbers of states.

5   SPACE-TIME BLOCK CODES

When the number of antennas is fixed, the decoding complexity of space-time trellis coding (measured by the number of trellis states at the decoder) increases exponentially as a function of the diversity level and transmission rate [36]. In addressing the issue of decoding complexity, Alamouti [38] discovered a remarkable space-time block coding scheme for transmission with two antennas. This scheme supports maximum likelihood detection based only on linear processing at the receiver. This scheme was later generalized in [39]

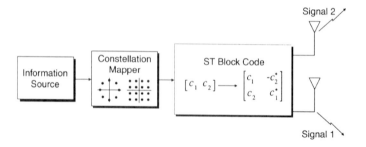

Figure 2.12   Transmitter Diversity with Space-Time Block Coding.

to an arbitrary number of antennas and is able to achieve the full diversity promised by the number of transmit and receive antennas. Here, we will briefly review the basics of space-time block codes. Figure 2.12 shows the baseband representation for space-time block coding with two antennas at the transmitter. The input symbols to the space-time block encoder are divided into groups of two symbols each. At a given symbol period, the two symbols in each group $\{c_1, c_2\}$ are transmitted simultaneously from the two antennas. The signal transmitted from antenna 1 is c_1 and the signal transmitted from antenna 2 is c_2. In the next symbol period, the signal $-c_2^*$ is transmitted from antenna 1 and the signal c_1^* is transmitted from antenna 2. Let h_1 and h_2 be the channels from the first and second transmit antennas to the receive antenna, respectively. The major assumption here is that h_1 and h_2 are constant over two consecutive symbol periods, that is

$$h_i(nT) = h_i((n+1)T), \qquad i = 1, 2$$

We assume a receiver with a single receive antenna, and we denote the received signals over two consecutive symbol periods as r_1 and r_2. The received signals can be written as:

$$r_1 = h_1 c_1 + h_2 c_2 + \eta_1 \qquad (2.12)$$
$$r_2 = -h_1 c_2^* + h_2 c_1^* + \eta_2 \qquad (2.13)$$

where η_1 and η_2 represent the AWGN and are modeled as i.i.d. complex Gaussian random variables with zero mean and power spectral density $N_o/2$ per dimension. We define the received signal vector $\mathbf{r} = [r_1 \ r_2^*]^T$, the code symbol vector $\mathbf{c} = [c_1 \ c_2]^T$, and the noise vector $\boldsymbol{\eta} = [\eta_1 \ \eta_2^*]^T$. Equations (2.12) and (2.13) can be rewritten in a matrix form as

$$\mathbf{r} = \mathbf{H} \cdot \mathbf{c} + \boldsymbol{\eta} \qquad (2.14)$$

where the channel matrix $\mathbf{H}$ is defined as

$$\mathbf{H} = \begin{bmatrix} h_1 & h_2 \\ h_2^* & -h_1^* \end{bmatrix} \tag{2.15}$$

The vector η is a complex Gaussian random vector with zero mean and covariance $N_o \cdot \mathbf{I}$. Let us define $\mathcal{C}$ as the set of all possible symbol pairs $\mathbf{c} = \{c_1, c_2\}$. Assuming that all symbol pairs are equiprobable, and since the noise vector η is assumed to be a multivariate AWGN, we can easily see that the optimum maximum likelihood decoder is

$$\hat{\mathbf{c}} = \arg\min_{\hat{\mathbf{c}} \in \mathcal{C}} \|\mathbf{r} - \mathbf{H} \cdot \hat{\mathbf{c}}\|^2 \tag{2.16}$$

The ML decoding rule in (2.16) can be further simplified by realizing that the channel matrix $\mathbf{H}$ is orthogonal and, hence, $\mathbf{H}^*\mathbf{H} = \rho \cdot \mathbf{I}$ where $\rho = |h_1|^2 + |h_2|^2$. Consider the modified signal vector $\tilde{\mathbf{r}}$ given by

$$\tilde{\mathbf{r}} = \mathbf{H}^* \cdot \mathbf{r} = \rho \cdot \mathbf{c} + \tilde{\eta} \tag{2.17}$$

where $\tilde{\eta} = \mathbf{H}^* \cdot \eta$. In this case the decoding rule becomes

$$\hat{\mathbf{c}} = \arg\min_{\hat{\mathbf{c}} \in \mathcal{C}} \|\tilde{\mathbf{r}} - \rho \cdot \hat{\mathbf{c}}\|^2 \tag{2.18}$$

Since $\mathbf{H}$ is orthogonal, we can easily verify that the noise vector $\tilde{\eta}$ will have a zero mean and covariance $\rho N_o \cdot \mathbf{I}$, i.e. the elements of $\tilde{\eta}$ are independent and identically distributed. Hence, it follows immediately that by using this simple linear combining, the decoding rule in (2.18) reduces to two separate, and much simpler, decoding rules for c_1 and c_2, as established in [38]. In fact, for the above 2×2 space-time block code, only two complex multiplications and one complex addition per symbol are required for decoding. Also, assuming that we are using a signaling constellation with 2^b constellation points, this linear combining reduces the number of decoding metrics that has to be computed for ML decoding from 2^{2b} to 2×2^b. It is also straight forward to verify that the SNR for c_1 and c_2 will be

$$\text{SNR} = \frac{\rho \cdot E_s}{N_o} \tag{2.19}$$

and hence a two branch diversity performance (i.e. a diversity gain of order two) is obtained at the receiver.

When the receiver uses M receive antennas, the received signal vector $\mathbf{r}_m$ at receive antenna m is

$$\mathbf{r}_m = \mathbf{H}_m \cdot \mathbf{c} + \eta_m \tag{2.20}$$

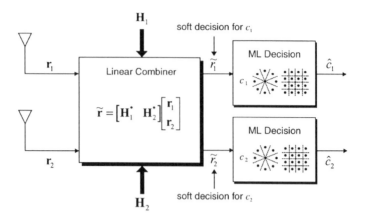

Figure 2.13   Receiver for Space-Time Block Coding.

where η_m is the noise vector and $\mathbf{H}_m$ is the channel matrix from the two transmit antennas to the mth receive antenna. In this case the optimum ML decoding rule is

$$\hat{\mathbf{c}} = \arg\min_{\hat{\mathbf{c}} \in \mathcal{C}} \sum_{m=1}^{M} \|\mathbf{r}_m - \mathbf{H}_m \cdot \hat{\mathbf{c}}\|^2 \qquad (2.21)$$

As before, in the case of M receive antennas, the decoding rule can be further simplified by pre-multiplying the received signal vector $\mathbf{r}_m$ by $\mathbf{H}_m^*$. In this case, the diversity order provided by this scheme is $2M$. Figure 2.13 shows a simplified block diagram for the receiver with two receive antennas. Note that the decision rule in (2.18) and (2.21) amounts to performing a hard decision on $\tilde{\mathbf{r}}$ and $\tilde{\mathbf{r}}_M = \sum_{m=1}^{M} \mathbf{H}_m^* \mathbf{r}_m$, respectively. Therefore, as shown in Figure 2.13, the received vector after linear combining, $\tilde{\mathbf{r}}_M$, can be considered as a soft decision for c_1 and c_2. When the space-time block code (STBC) is concatenated with an outer conventional channel code, like a convolutional code, these soft decisions can be fed to the outer channel decoder to yield a better performance. Note also that for the above 2×2 STBC, the transmission rate is 1 while achieving the maximum diversity gain possible with two transmit antennas.

The extension of the above STBC was studied in [39]. A general technique was developed for constructing space-time block codes for $N > 2$ that provide the maximum diversity promised by the number of transmit and receive antennas. These codes retain the simple ML decoding algorithm based on only linear processing at the receiver [38]. It was also shown that for real signal constellations (PAM constellation), space-time block codes with transmission rate 1 can be constructed [39]. However, for a general complex constellations like M-QAM or M-PSK, it *is not known* whether a space-time block code with transmission rate 1 and simple linear processing that will give the maximum diversity gain with $N > 2$ transmit antennas **does exist or not**. Moreover, it was also shown that such code where the number of transmit antennas N equals the number of equals both the number of information symbols transmitted and the number of time slots need to transmit the code block **does not exit**. However for rates < 1, such codes can be found. For example, assuming that the transmitter unit uses 4 transmit antennas, a rate 4/8 (i.e. it is a rate $1/2$) space-time block code is given by

$$
\begin{bmatrix} c_1 \\ c_2 \\ c_3 \\ c_4 \end{bmatrix} \rightarrow
\begin{bmatrix}
c_1 & -c_2 & -c_3 & -c_4 & c_1^* & -c_2^* & -c_3^* & -c_4^* \\
c_2 & c_1 & c_4 & -c_3 & c_2^* & c_1^* & c_4^* & -c_3^* \\
c_3 & -c_4 & c_1 & c_2 & c_3^* & -c_4^* & c_1^* & c_2^* \\
c_4 & c_3 & -c_2 & c_1 & c_4^* & c_3^* & -c_2^* & c_1^*
\end{bmatrix}
\tag{2.22}
$$

In this case, at time $t = 1$, c_1, c_2, c_3, c_4 are transmitted from antennas 1 through 4, respectively. At time $t = 2$, $-c_2, c_1, -c_4, c_3$ are transmitted from antenna 1 through 4, respectively, and so on. For this example, let $r_1, r_2, \cdots, r_8$ be the received signals at time $t = 1, 2, \cdots, 8$ respectively. Define the new received signal vector $\mathbf{r} = [r_1, r_2, r_3, r_4, r_5^*, r_6^*, r_7^*, r_8^*]^T$. In this case we can write the received signal vector $\mathbf{r}$ at the receive antenna as

$$
\mathbf{r} = \mathbf{H} \cdot \mathbf{c} + \boldsymbol{\eta}
\tag{2.23}
$$

where $\boldsymbol{\eta}$ is an 8×1 AWGN noise vector and $\mathbf{H}$ is 8×4 channel matrix given by

$$
\mathbf{H} =
\begin{bmatrix}
h_1 & h_2 & h_3 & h_4 \\
h_2 & -h_1 & h_4 & -h_3 \\
h_3 & -h_4 & -h_1 & h_2 \\
h_4 & h_3 & -h_2 & -h_1 \\
h_1^* & h_2^* & h_3^* & h_4^* \\
h_2^* & -h_1^* & h_4^* & -h_3^* \\
h_3^* & -h_4^* & -h_1^* & h_2^* \\
h_4^* & h_3^* & -h_2^* & -h_1^*
\end{bmatrix}
\tag{2.24}
$$

We can immediately see that $\mathbf{H}$ is orthogonal, that is $\mathbf{H}^*\mathbf{H} = \rho_4 \cdot \mathbf{I}$, where $\rho_4 = 2 \cdot \sum_{i=1}^4 |h_i|^2$. Therefore, the same procedure used for decoding the simple 2×2 STBC can be used for this code too. In this case, the SNR for $c_1, \cdots, c_4$ is $\rho_4 E_s / N_o$, i.e. a 4-branch diversity performance + 3 dB coding gain is achieved. The 3 dB coding gain comes from the (intuitive) fact that 8 time slots are used to transmit 4 information symbols.

Note that the decoding of ST block codes requires knowledge of the channel at the receiver. The channel state information can be obtained at the receiver by sending training or pilot symbols or sequences to estimate the channel from each of the transmit antennas to the receive antenna [47, 48, 49, 50, 51, 52, 53, 54]. For one transmit antenna, there exist differential detection schemes, such as DPSK, that neither require knowledge of the channel nor employ pilot or training symbol transmission. These differential decoding schemes are used, for example, in the IS-54 cellular standard ($\pi/4$-DPSK). This motivates the generalization of differential detection schemes for the case of multiple transmit antennas. A partial solution to this problem was proposed in [42] for the 2×2 code, where it was assumed that the channel is not known at the receiver. In this scheme, the detected pair of symbols at time $t - 1$ are used to estimate the channel at the receiver and these channel estimates are used for detecting the pair of symbols at time t. However, the scheme in [42] requires the transmission of known pilot symbols at the beginning and hence is not fully differential. The scheme in [42] can be thought as a joint data channel estimation approach which can lead to error propagation. In [41], a true differential detection scheme for the 2×2 code was constructed. This scheme shares many of the desirable properties of DPSK: it can be demodulated with or without CSI at the receiver, achieves full diversity gain in both cases, and there is a simple noncoherent receiver that performs within 3 dB of the coherent receiver. However, this scheme has some limitations. First, the encoding scheme expands the signal constellation for non-binary signals. Second, it is limited only to the $N = 2$ space-time block code for complex constellations and to the case $N \le 8$ for real constellations. This is based on the results in [39] that the 2×2 STBC is an orthogonal design and complex orthogonal designs do not exist for $N > 2$. In [55], another approach for differential modulation with transmit diversity based on group codes was proposed. This approach can be applied to any number of antennas and to any constellation. The group structure of theses codes greatly simplifies the analysis of these schemes, and may also yield simpler and more transparent modulation and demodulation procedures. A different no-differential approach to transmit diversity when the channel is not known at the receiver is reported in [56, 57] but this approach requires exponential encoding and decoding complexities.

6   INTERFERENCE SUPPRESSION WITH SPACE-TIME BLOCK CODES

The properties of the space-time block coding scheme in [38] and its extension in [39] can be further exploited to develop efficient interference suppression techniques that can be used to increase system capacity or increase throughput for individual users. In general, we consider a multiuser environment with K synchronous co-channel users where each user is equipped with N transmit antennas and uses a STBC with N transmit antenna. In general, in this scenario, there will be $K \times N$ interfering signals arriving at the receiver.

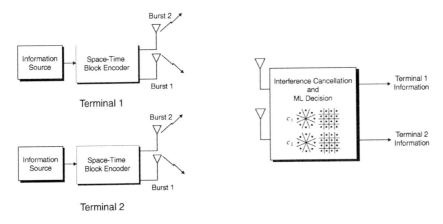

Figure 2.14   Interference Cancellation with Space-Time Block Codes.

Therefore, classical interference suppression techniques [58] with multiple receive antennas will require $N \times (K - 1) + 1$ antennas at the receiver in order to suppress signals from the $K - 1$ co-channel space-time users and achieve a diversity order of N for the desired terminal. By exploiting the temporal and spatial structure space-time block codes, it can be shown [43, 44, 59] that only K antennas are required to suppress the interference from the $K - 1$ co-channel users while maintaining the diversity order of N provided by the space-time block code. Given the assumption that the receiver is equipped with $M \geq K$ antennas, zero forcing (ZF) and minimum mean-squared error (MMSE) interference suppression techniques that exploit the structure of the STBC are developed in [43, 59]. These techniques will *perfectly* suppress the interference from the $K - 1$ co-channel users and provide a diversity order of $N \times (M - K + 1)$ while maintaining the simple linear processing feature of the space-time block codes.

We outline these interference cancellation schemes for the 2×2 case here. For a more detailed treatment the reader is referred to [43]. Figure 2.14 shows a simple scenario for two synchronous co-channel space-time users (each employs the 2×2 STBC) and a receiver with two receiver antennas. Using the signal model developed above, the received signal vectors at antennas 1 and 2 are

$$\mathbf{r}_1 = \mathbf{H}_1 \cdot \mathbf{c} + \mathbf{G}_1 \cdot \mathbf{s} + \boldsymbol{\eta}_1 \qquad (2.25)$$

$$\mathbf{r}_2 = \mathbf{H}_2 \cdot \mathbf{c} + \mathbf{G}_2 \cdot \mathbf{s} + \boldsymbol{\eta}_2 \qquad (2.26)$$

where $\mathbf{r}_1$ is the received vector at antenna 1, $\mathbf{r}_2$ is the received vector at antenna 2, $\mathbf{c}$ is the vector of code symbols from first user, and $\mathbf{s}$ is the vector of code

Table 2.1 Zero forcing IC and ML decoding algorithm for space-time block codes.

$(\mathbf{c}, \Delta) = \text{ZF.DEC}(\mathbf{r}_1, \mathbf{r}_2, \mathbf{H}_1, \mathbf{H}_2, \mathbf{G}_1, \mathbf{G}_2)$
{

$$\begin{aligned}
\tilde{\mathbf{r}} &= \mathbf{r}_1 - \mathbf{G}_1 \mathbf{G}_2^{-1} \mathbf{r}_2 \\
\tilde{\mathbf{H}} &= \mathbf{H}_1 - \mathbf{G}_1 \mathbf{G}_2^{-1} \mathbf{H}_2 \\
\sigma^2 &= 1 + \text{Tr}\{\mathbf{G}_1 \mathbf{G}_1^*\} / \text{Tr}\{\mathbf{G}_2 \mathbf{G}_2^*\} \\
\hat{\mathbf{c}} &= \arg\min_{\hat{\mathbf{c}} \in \mathcal{C}} \left\| \sigma^{-1} \left(\tilde{\mathbf{r}} - \tilde{\mathbf{H}} \cdot \hat{\mathbf{c}} \right) \right\|^2 \\
\Delta &= \left\| \sigma^{-1} \left(\tilde{\mathbf{r}} - \tilde{\mathbf{H}} \cdot \hat{\mathbf{c}} \right) \right\|^2
\end{aligned}$$

}

symbols from second user. The matrices $\mathbf{H}_1$ and $\mathbf{H}_2$ are the channel matrices from the first space-time user to the first and second receive antennas, respectively, and are defined similar to (2.15). Similarly, the matrices $\mathbf{G}_1$ and $\mathbf{G}_2$ are the channel matrices from the second space-time user to the first and second receive antennas, respectively. The last two equations can be rewritten as

$$\begin{aligned}
\mathbf{r} &= \begin{bmatrix} \mathbf{r}_1 \\ \mathbf{r}_2 \end{bmatrix} = \mathbf{H} \cdot \tilde{\mathbf{c}} + \eta \\
&= \begin{bmatrix} \mathbf{H}_1 & \mathbf{G}_1 \\ \mathbf{H}_2 & \mathbf{G}_2 \end{bmatrix} \begin{bmatrix} \mathbf{c} \\ \mathbf{s} \end{bmatrix} + \begin{bmatrix} \eta_1 \\ \eta_2 \end{bmatrix}
\end{aligned} \tag{2.27}$$

In the zero-forcing solution, the interference between the two space-time co-channel users is removed, without any regard to noise enhancement, by using a matrix linear combiner $\mathbf{W}$ such that

$$\mathbf{W} \cdot \mathbf{r} = \begin{bmatrix} \tilde{\mathbf{r}}_1 \\ \tilde{\mathbf{r}}_2 \end{bmatrix} = \begin{bmatrix} \tilde{\mathbf{H}} & 0 \\ 0 & \tilde{\mathbf{G}} \end{bmatrix} \begin{bmatrix} \mathbf{c} \\ \mathbf{s} \end{bmatrix} + \begin{bmatrix} \tilde{\eta}_1 \\ \tilde{\eta}_2 \end{bmatrix} \tag{2.28}$$

In this case, the modified received signal vector $\tilde{\mathbf{r}}_1$ depends only on signals from first terminal and the modified received signal vector $\tilde{\mathbf{r}}_2$ depends only on signals from second terminal. It was shown in [43] that a solution for $\mathbf{W}$ is given by

$$\mathbf{W} = \begin{bmatrix} \mathbf{I}_2 & -\mathbf{G}_1 \mathbf{G}_2^{-1} \\ -\mathbf{H}_2 \mathbf{H}_1^{-1} & \mathbf{I}_2 \end{bmatrix} \tag{2.29}$$

Table 2.2   MMSE IC decoding algorithm for space-time block codes.

$(\mathbf{c}, \Delta) = \text{MMSE.DEC}(\mathbf{r}_1, \mathbf{r}_2, \mathbf{H}_1, \mathbf{H}_2, \mathbf{G}_1, \mathbf{G}_2, \Gamma)$
{

$$\tilde{\mathbf{r}} = [\mathbf{r}_1^T \mathbf{r}_2^T]^T$$

$$\tilde{\mathbf{H}} = \begin{bmatrix} \mathbf{H}_1 & \mathbf{G}_1 \\ \mathbf{H}_2 & \mathbf{G}_2 \end{bmatrix}$$

$$\mathbf{M} = \mathbf{H}\mathbf{H}^* + \frac{1}{\Gamma}\mathbf{I}_4$$

$$\mathbf{h}_1 = [h_{11}\ h_{21}^*\ h_{12}\ h_{22}^*]^T$$
$$= \text{first column of } \mathbf{H}$$

$$\mathbf{h}_2 = [h_{21}\ -h_{11}^*\ h_{22}\ -h_{12}^*]^T$$
$$= \text{second column of } \mathbf{H}$$

$$\boldsymbol{w}_1 = \mathbf{M}^{-1}\mathbf{h}_1$$

$$\boldsymbol{w}_2 = \mathbf{M}^{-1}\mathbf{h}_2$$

$$\hat{\mathbf{c}} = \underset{\hat{c} \in \mathcal{C}}{\arg\min} \left\{ \|\boldsymbol{w}_1^*\mathbf{r} - \hat{c}_1\|^2 + \|\boldsymbol{w}_2^*\mathbf{r} - \hat{c}_2\|^2 \right\}$$

$$\Delta = \|\boldsymbol{w}_1^*\mathbf{r} - \hat{c}_1\|^2 + \|\boldsymbol{w}_2^*\mathbf{r} - \hat{c}_2\|^2$$

}

It is interesting to note that by using this matrix linear combiner $\mathbf{W}$, the matrices $\tilde{\mathbf{H}}$ and $\tilde{\mathbf{G}}$ will have the same structure as that of the channel matrix $\mathbf{H}$ in (2.15). Hence, using the matrix linear combiner in (2.29) will reduce the problem of detecting the two co-channel space-time users into two separate problems that have a much simpler solution as pointed out before. Table 2.1 shows the algorithm for the zero-forcing interference cancellation and maximum likelihood decoding of STBC. In the MMSE interference suppression technique, let us assume, for example, that we are interested in decoding signals from the first space-time user. In this case, the receiver selects two linear combiners $\boldsymbol{w}_1$ and $\boldsymbol{w}_2$ such that

$$J_1(\boldsymbol{w}_1) = \|\boldsymbol{w}_1^*\mathbf{r} - c_1\|^2 \qquad \text{and} \qquad J_2(\boldsymbol{w}_2) = \|\boldsymbol{w}_2^*\mathbf{r} - c_2\|^2 \quad (2.30)$$

are minimized. It was shown in [43] that the optimum solution is given by

$$\boldsymbol{w}_1 = \mathbf{M}^{-1}\mathbf{h}_1 \qquad \text{and} \qquad \boldsymbol{w}_2 = \mathbf{M}^{-1}\mathbf{h}_2 \qquad (2.31)$$

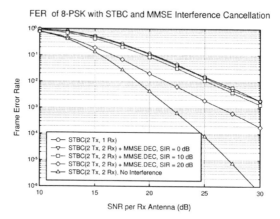

Figure 2.15   MMSE Interference Cancellation with Space-Time Block Codes.

where $\mathbf{M} = \mathbf{HH}^* + \frac{1}{\Gamma}\mathbf{I}$, $\Gamma = E_s/N_o$ is the signal to noise ratio (SNR), $\mathbf{h}_1 = [h_{11}\ h_{21}^*\ h_{12}\ h_{22}^*]^T$ is the first column of $\mathbf{H}$, and $\mathbf{h}_2 = [h_{21}\ -h_{11}^*\ h_{22}\ -h_{12}^*]^T$ is the second column of $\mathbf{H}$. It was shown in [43] that w_1 and w_2 are orthogonal, and hence, errors in decoding c_1 do not affect decoding c_2 and visa versa, thereby maintaining the separate detection feature for STBC decoding. Note that the MMSE solution will reduce to the ZF solution outlined earlier as $\Gamma \to \infty$. Table 2.2 outlines the algorithm description for MMSE interference suppression and decoding of STBC. For a more detailed treatment of both the ZF and MMSE solutions the reader is referred to [43].

Figure 2.15 shows the performance of the MMSE interference cancellation scheme as a function of SNR and signal to interference ratio (SIR) for two co-channel space-time users each using the 2×2 space-time block code and a receiver with 2 receive antennas. Note that the performance of the ZF interference cancellation will always be the same as that of a single space-time user with one receiver antenna.

7   APPLICATIONS OF SPACE-TIME CODING TO WIRELESS

As pointed out earlier, one of the goals of the third and fourth generation wireless systems is to provide broadband access to both mobile and stationary users. Real-time multi-media services (such as video conferencing) would

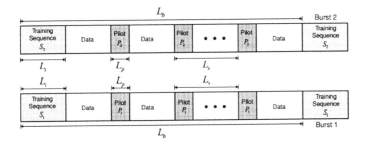

Figure 2.16 Downlink Slot structure for STCM-based modem.

require data rates 2-3 orders of magnitude larger than what is offered by current wireless technologies. A higher spectral efficiency can be achieved by using multiple transmit and/or receive antennas [1, 19, 60]. Space-time coding techniques with multiple transmit antennas offer the best possible trade-off between power consumption and spectral efficiency in multipath radio channels. Space-time coding and signal processing techniques with multiple transmit antennas have been recently adopted in third generation cellular standard (e.g. cdma2000 [61] and W-CDMA [62]) and also have been proposed for wireless local loop applications (Lucent's BLAST project [6]) and wide-area packet data access (AT&T's Advanced Cellular Internet Service [5]). In this section we will outline several examples of application of space-time coding to different wireless applications.

7.1 APPLICATION TO NARROW BAND TDMA CELLULAR

In this Section, we will present a general architecture for a narrow band TDMA modem with space-time coding and 2 transmit antennas [47]. For brevity, we will present the modem architecture for the downlink only. The uplink modem will have a similar architecture, except that the framing and timing structure will be different and must allow for a guard time between different asynchronous (due to difference in propagation delay) bursts from different users. The system architecture that we propose is similar, but not identical, to that of the IS-136 US cellular standard. Figure 2.16 shows the basic TDMA time slot structure, employing a signaling format which interleaves training and synchronization sequences, pilot sequences, and data is used. In each TDMA slot, two bursts are transmitted, one from each antenna. The training sequences S_1 and S_2 will be used for timing and frequency synchronization at the receiver. In addition, the transmitter inserts periodic and orthogonal pilot sequences P_1 and P_2 which are used, along with the training sequences S_1 and

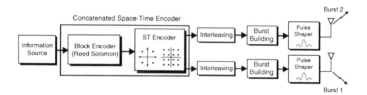

Figure 2.17   Base station transmitter with STCM and 2 transmit antennas.

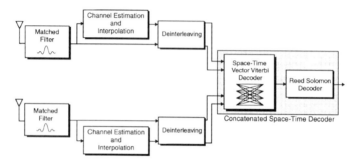

Figure 2.18   Mobile receiver with STCM and 2 receive antennas.

S_2, at the receiver to estimate the channel from each of the transmit antennas to the corresponding receive antenna. Figure 2.17 shows a block diagram for the transmitter, where in addition to the space-time encoder, a high rate Reed Solomon (RS) block encoder is used as an outer code. The RS outer code is used to correct the few symbol errors at the output of the space-time decoder. The output of the RS encoder is then encoded by a space-time channel encoder and the output of the space-time encoder is split into 2 streams of encoded modulation symbols. Each stream of encoded symbols is then independently interleaved using a block symbol-by-symbol interleaver. The transmitter inserts the corresponding training and periodic pilot sequences in each of the two bursts. Each burst is then pulse-shaped and transmitted from the corresponding antenna. The signal transmitted from the ith antenna, $i = 1, 2$, can be written as

$$s_i(t) = \sqrt{E_s} \cdot \sum_l c_i(l) p(t - lT_s) \tag{2.32}$$

where $T_s = 1/R_s$ is the symbol period and $p(t)$ is the transmit filter pulse. Figure 2.18 shows the corresponding block diagram of a mobile receiver equipped with 2 receive antennas. After down conversion to baseband, the received signal at each antenna element is filtered using a receive filter with impulse

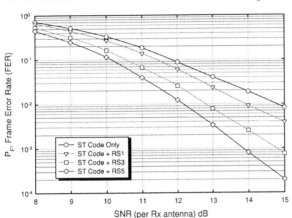

Figure 2.19 FER Performance of 8-PSK 32-state space-time code with 2 transmit and 2 receive antennas at 180 Hz Doppler with different coding rates.

response $\bar{p}(t)$ that is matched to the transmit pulse shape $p(t)$. The output of the matched filters is oversampled at a rate that is an integer multiple of the symbol rate. Received samples corresponding to the training sequences S_1 and S_2 are used for timing and frequency synchronization. The received samples at the optimum sampling instant are then split into two streams. The first one contains the received samples corresponding to the pilot and training symbols. These are used to estimate the corresponding CSI $\hat{\mathcal{H}}(l)$ at the pilot and training sequence symbols. The receiver then uses an appropriately designed interpolation filter to interpolate those trained CSI estimates and obtain accurate interpolated CSI estimates for the whole burst. The second stream contains the received samples corresponding to the superimposed information symbols. The interpolated CSI estimates along with the received samples corresponding to the information symbols are then deinterleaved using a block symbol-by-symbol deinterleaver and passed to a vector maximum likelihood sequence decoder followed by a RS decoder.

Figure 2.19 shows the performance of the above modem architecture with a 32-state 8-PSK space-time code with two transmit and two receive antennas, and with different coding rate options (see [47] for details). At 10 Hz Doppler, this modem architecture with the 32-state 8-PSK STC would be able to deliver almost 56 kbps (over a 30 kHz bandwidth) with 10% frame error rate at 18 dB, and 11 dB for 1 and 2 receive antennas, respectively. At 180 Hz Doppler, the

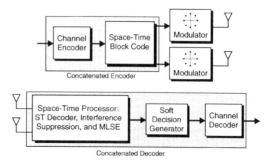

Figure 2.20   Concatenation of Space-Time Block Code and Conventional Channel Coding Schemes for Increasing Capacity.

required SNR would be 20 dB and 12 dB respectively. These results assume basic IS-136 channelization and framing structure. As pointed out in [47], this architecture has the potential of almost doubling the current data rates supported by the 1S-136 cellular standard.

7.2   APPLICATIONS TO INCREASING CAPACITY/THROUGHPUT OF WIRELESS SYSTEMS

First, we consider a scenario where K *synchronized* terminal units each with two transmit antennas communicate with a base-station having $M \geq K$ receive antennas. Increased system capacity (in terms of the number co-channel terminals that can simultaneously communicate with the base-station) can be attained while providing diversity benefits to each terminal by using a concatenated coding scheme where the inner code is a space-time block code and the outer code is a conventional channel error correcting code (a TCM, a convolutional code, or a RS code, for example), as shown in Figure 2.20. More specifically, information symbols are first encoded using a conventional channel code. The output of this channel code is then encoded using a space-time block encoder with two transmit antennas (N transmit antennas in general can be used with the appropriate space-time block code). At the receiver, the inner space-time block code is used to suppress interference from the other co-channel terminals using, for example, the MMSE interference suppression technique described above. In the above technique, a hard decision is applied on the output of the interference canceler to produce an estimate for the transmitted information symbols. That is, given the two IC weight vectors w_{i1} and w_{i2} corresponding to some terminal i, the receiver forms the two decision variables

$$\xi_{i1} = w_{i1}^* r \qquad (2.33)$$
$$\xi_{i2} := w_{i2}^* r \qquad (2.34)$$

A hard decision is then applied on these decision variables to decode the two transmitted symbols corresponding to the i-th terminal. However, in the case when the space-time code is concatenated with an outer conventional channel code, the decision variables ξ_1 and ξ_2 are used as soft decisions for the transmitted information symbols and then fed to the conventional channel decoder. This will improve the error rate performance of the conventional channel code as compared to using hard decision. Thus, in this scheme, we are using the structure of the inner space-time code for interference suppression and we are able to support many co-channel terminals while providing diversity benefit to those terminals. At the same time, the inner space-time decoder provides soft decision output for the outer conventional channel code which will provide protection against channel errors.

The above space-time block coding and MMSE interference suppression technique can also be used in situations where increasing the data rate or the data throughput is of interest. In this case, information symbols from a transmitting terminal are split into L parallel streams. Stream l is then encoded using a conventional channel code with rate R_l and then encoded with a space-time block encoder with two transmitting antennas (as before, N transmit antennas in general can be used with the appropriate space-time block codes). The coding rates for each of the L parallel streams are chosen such that $R_1 < R_2 < \cdots < R_L$. In this case, symbols transmitted in stream l will have better immunity against channel errors than symbols transmitted in stream u where $u > l$. The base station receiver is assumed to be equipped with L receive antennas. The base station receiver treats each stream as a different user and uses the above MMSE interference suppression technique to generate soft decisions ξ_{11} and ξ_{12} for the data in the first stream. These soft decision are then fed into the decoder corresponding to the first channel code. The output information symbols are then *re-encoded* with the same channel code for the first stream. Since the first stream has the smallest coding rate R_1, it will have the best immunity against channel errors and most likely it will be error free. The resulting symbols are then used to subtract the contributions of the first stream in the received signal while decoding the remaining $L - 1$ streams. In decoding the remaining $L - 1$ streams , the decoder will decode signals from the second stream first since it will have the best immunity against channel errors among the remaining $L - 1$ streams. Then the receiver cancels out the contribution of the second stream in the received signal. This process is repeated until all streams are decoded. In this case, we define the system throughput as

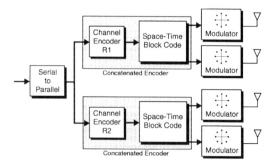

Figure 2.21   Parallel Transmission with Space-Time Block Coding for Increased System Throughput.

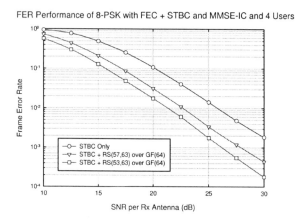

Figure 2.22   FER Performance of 4 co-channel space-time users with a concatenated coding scheme over a flat fading channel.

$$\rho = \frac{1}{L} \sum_{l=1}^{L} R_l \cdot (1 - FER_l) \qquad (2.35)$$

where FER_l is the frame error rate of stream l. As we will see from the simulation, this will increase the system throughput at low signal to noise ratios. Figure 2.21 shows a block diagram for a terminal that uses 4 transmit antennas. In this case, the input information stream is split into two parallel streams *i.e.* $(L = 2)$.

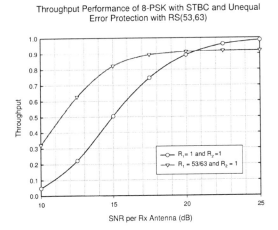

Figure 2.23 Throughput Performance of Parallel Transmission with Space-Time Block Coding with Unequal Coding.

Figure 2.22 shows the performance of the system in Figure 2.20 where a concatenated coding scheme is used. The figure show the FER of any of the 4 users with different coding rates. There was 4 co-channel users and each uses the 2×2 STBC and the receiver had 4 antennas. The above MMSE-IC scheme was used to separate the 4 users. This scheme is suitable for fixed wireless access applications. Figure 2.23 shows the throughput performance of the system in 2.21. Combined MMSE interference cancellation and decoding of the STBC was used to separate the two different data streams. Using this parallel transmission and making use of the STBC properties to separate the two streams will allow for doubling the data rate. Also if one of the two data streams is coded heavier than the other one, increased throughput can be obtained especially at low SNR. See [45] for more details.

7.3 APPLICATION TO BROADBAND WIRELESS

Figures 2.24 and 2.25 show simplified block diagrams for the transmitter and receiver, respectively, for an OFDM modem with a concatenated space-time coding scheme. This architecture [63] is suitable for broadband wireless communications applications (Similar work, but based on space-time block codes, can be found in [64, 65]). The input information symbols are first encoded by an outer conventional channel code. The output of the outer code is then space-time encoded. Each stream of the space-time code output streams

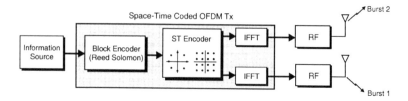

Figure 2.24   Transmitter for Space-Time Coded OFDM for Broadband Applications.

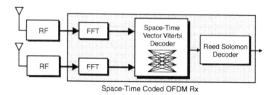

Figure 2.25   Receiver for Space-Time Coded OFDM for Broadband Applications.

is then OFDM modulated and send over the corresponding antenna. At the receiver, the signal at each receive antenna is OFDM-demodulated. The demodulated signals from antennas are then fed into the space-time decoder followed by the outer decoder. Figure 2.26 shows the simulation results for the above OFDM space-time coded modem. In this simulation, the available bandwidth is 1 MHz and the maximum Doppler frequency is 200 Hz. The number of OFDM tones used for modulation is 256. These correspond to a subcarrier separation of 3.9 KHz and OFDM frame duration of 256 μs. A cyclic prefix of 40 μs duration is added to each frame. Each tones modulates a 4-PSK constellation, although higher order M-PSK or M-QAM may be used. We used a 16-state 4-PSK space-time code [36] with 2 transmit and 2 receive antennas together with an outer (72,64,9) RS code over GF(2^7). We plot the frame error probability as function of SNR for different delay spreads in Fig. 2.26. From this plot, we can see that an E_b/N_o between 2.7-4 dB (depending on the delay spread) is needed to achieve a data rate of 1.5 Mbps. This technique can be used also with the combined space-time block coding and interference suppression scheme, as shown in Figure 2.27 to yield even higher data rates (multiples of Mbps / 1 MHz) over a wireless channel.

8   CONCLUSIONS

Space-Time coding is a new coding/signal processing framework for wireless communication systems with multiple transmit and multiple receive antennas. This new framework has the potential of dramatically improving the

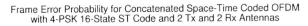

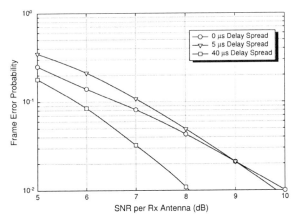

Figure 2.26 FER of Concatenated Space-Time Coded OFDM with 4-PSK 16-State STC with 2Tx and 2Rx Antennas.

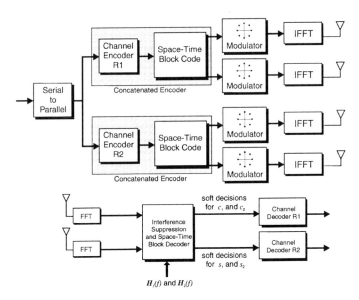

Figure 2.27 Parallel Transmission with Space-Time Block Coding for Increased System Throughput over Delay Spread Channels.

capacity and data rates. In addition, this framework presents the best tradeoff between spectral efficiency and power consumption. Space-Time codes (designed so far) come in two different flavors. Space-Time trellis codes offer the maximum possible diversity gain and a coding gain without any sacrifice in the transmission bandwidth. The decoding of these codes, however, would require the use of a vector form of the Viterbi decoder. Space-Time block codes, however, offer a much simpler way of obtaining transmit diversity without any sacrifice in bandwidth and without requiring huge decoding complexity. In fact, the structure of of the space-time block codes is such that it allows for very simple signal processing (linear combining) for encoding/decoding, differential encoding/detection, and interference cancellation. This new signal processing framework offered by space-time codes can be used to enhance the data rate and/or capacity in various wireless applications. That is the reason many of these space-time coding ideas have already found their way to some of the current 3rd generation wireless systems standards.

References

[1] G. J. Foschini and M. J. Gans, "On Limits of Wireless Communications in a Fading Environment when Using Multiple Antennas," *Wireless Communications Magazine*, vol. 6, pp. 311–335, March 1998.

[2] E. Telatar, "Capacity of Multi-Antenna Gaussian Channels," technical memorandum, AT&T Bell Laboratories, June 1995.

[3] N. Sollenberger and S. Kasturia, "Evolution of TDMA (IS-54/IS-136) to Foster Further Growth of PCS," in *Proc. ICUPC, International Conference on Universal Personal Communications,1996*, (Boston,MA), 1996.

[4] "Special Issue on the European Path Towards UMTS," *IEEE Personal Communications Magazine*, vol. 2, February 1995.

[5] L. J. Cimini, Jr. , J. C.-I. Chuang, and N. R. Sollenberger, " Advanced Cellular Inernet Service (ACIS)," *IEEE Communications Magazine*, vol. 36, pp. 150–159, October 1998.

[6] P. W. Wolniansky, G. J. Foschini, G. D. Golden, and R. A. Valenzuela, "V-BLAST: An Architecture for Realizing Very High Data Rates over Rich Scattering Wireless Channels," in *Proc. ISSSE-98*, pp. 295–300, Sep. 1998.

[7] D. J. Goodman, "Trends in Cellular and Cordless Communications," *IEEE Communications Magazine*, vol. 29, pp. 31–40, June 1991.

[8] J. H. Winters, "Optimum Combining in Digital Mobile Radio with Cochannel Interference," *IEEE J. Select. Areas Commun.*, vol. JSAC-2(4), pp. 528–539, July 1984.

[9] J. H. Winters, "Optimum Combining for Indoor Radio Systems with Multiple Users," *IEEE Trans. Commun.*, vol. COM-35(11), pp. 1222–1230, November 1987.

[10] J. H. Winters, "On the Capacity of Radio Communication Systems with Diversity in a Rayleigh Fading Environment," *IEEE J. Select. Areas Commun.*, vol. JSAC-5(5), pp. 871–878, June 1987.

[11] P. Balaban and J. Salz, "Optimum Diversity Combining and Equalization in Digital Data Transmission with Application to Cellular Mobile Radio," *IEEE Trans. Veh. Tech.*, vol. VT-40(2), pp. 342–354, May 1991.

[12] P. Balaban and J. Salz, "Optimum Diversity Combining and Equalization in Data Transmission with Application to Cellular Mobile Radio - Part I: Theoretical Considerations," *IEEE Trans. Commun.*, vol. COM-40(5), pp. 885–894, May 1992.

[13] P. Balaban and J. Salz, "Optimum Diversity Combining and Equalization in Data Transmission with Application to Cellular Mobile Radio - Part II:Numerical results," *IEEE Trans. Commun.*, vol. COM-40(5), pp. 895–907, May 1992.

[14] J. G. Proakis, "Adaptive Equalization for TDMA Digital Mobile Radio," *IEEE Trans. Veh. Technol.*, vol. VT-40(2), pp. 333–341, May 1991.

[15] R. Price and J. P.E. Green, "A Communication Technique for Multipath Channels," *Proc. IRE*, vol. 46, pp. 555–570, March 1958.

[16] G. Turin, "Introduction to Spread-Spectrum Antimultipath Techniques and their Application to Urban Digital Radio," *Proc. of IEEE*, vol. 68, pp. 328–353, March 1980.

[17] F. Lotse, J.-E. Berg, U. Forssen, and P.Idhal, "Base station polarization diversity reception in macrocellular systems at 1800 MHz," in *Proc. Vehicular Technology Conference, 1996*, vol. 3, pp. 1643–1646, May 1996.

[18] W. C. Jakes, *Microwave Mobile Communications*. New York: John Wiley and Sons, 1974.

[19] G. Foschini, "Layered Space-Time Architecture for Wireless Communication in a Fading Environment when Using Multi-element antennas," *Bell Labs Technical Journal*, vol. 1, pp. 41–59, Autumn 1996.

[20] A. Narula, M.Trott, and G. Wornell, "Performance Limits of Coded Diversity Methods for Transmitter Antenna Arrays," *IEEE Trans. Information Theory*, vol. 45, pp. 2418 – 2433, November 1999.

[21] P. S. Henry and B. S. Glance, "A New Approach to High Capacity Digital Mobile Radio," *Bell Syst. Tech. Journal*, vol. 51, pp. 1611–1630, September 1972.

[22] J. H. Winters, "Switched Diversity with Feedback for DPSK Mobile Radio Systems," *IEEE Trans. on Vehicular Technology*, vol. VT-32, pp. 134–150, February 1983.

[23] A. Wittneben, "Base Station Modulation Diversity for Digital SIMUL-CAST," in *Proc. IEEE VTC'91*, vol. 1, (St Louis, USA), pp. 848–853, 1991.

[24] N. Seshadri and J. H. Winters, "Two Schemes for Improving the Performance of Frequency-Division Duplex (FDD) Transmission Systems Using Transmitter Antenna Diversity," *International Journal of Wireless Information Networks*, vol. 1, pp. 49–60, Jan 1994.

[25] A. Wittneben, "A New Bandwidth Efficient Transmit Antenna Modulation Diversity Scheme for Linear Digital Modulation," in *Proc. IEEE ICC'93*, vol. 3, (Geneva, Switzerland), pp. 1630–1634, 1993.

[26] J.-C. Guey, M. P. Fitz, M. R. Bell, and W.-Y. Kuo, "Signal Design for Transmitter Diversity Wireless Communication Systems over Rayleigh Fading Channels," in *Proc. IEEE VTC'96*, vol. 1, (Atlanta, GA), pp. 136–140, 1996.

[27] J. H. Winters, "Diversity gain of transmit diversity in wireless systems with Rayleigh fading," in *Proc. IEEE ICC'94*, vol. 2, (New Orleans,LA), pp. 1121–1125, 1994.

[28] J. H. Winters, "Diversity gain of transmit diversity in wireless systems with Rayleigh fading," *IEEE Trans. Veh. Technol.*, vol. 47, pp. 119–123, February 1998.

[29] A. Hiroike, F. Adachi, and N. Nakajima, "Combined Effects of Phase Sweeping Transmitter Diversity and Channel Coding," *IEEE Trans. Veh. Technol.*, vol. VT-41, pp. 170–176, May 92.

[30] T. Hattori and K. Hirade, "Multitransmitter Simulcast Digital signal Transmission by Using Frequency Offset Strategy in Land Mobile Radio-Telephone," *IEEE. Trans. Veh. Technol.*, vol. VT-27, pp. 231–238, 1978.

[31] V. Weerackody, "Diversity for the Direct-Sequence Spread Spectrum System Using Multiple Transmit Antennas," in *Proc. ICC'93*, vol. III, (Geneva, Switzerland), pp. 1503–1506, May 1993.

[32] N. Seshadri, V. Tarokh, and A. R. Calderbank, "Space-Time Codes for High Data Rate Wireless Communications: Code Construction," in *Proc. IEEE VTC'97*, vol. 2, (Phoenix,AZ), pp. 637–641, 1997.

[33] V. Tarokh, N. Seshadri, and A. R. Calderbank, "Space-Time Codes for High Data Rate Wireless Communications: Performance Criterion and Code Construction," in *Proc. IEEE ICC'97*, vol. 1, (Montreal,Canada), pp. 299–303, 1997.

[34] V. Tarokh, A. F. Naguib, N. Seshadri, and A. R. Calderbank, "Space-Time Codes for High Data Rate Wireless Communications: Mismatch Analysis," in *Proc. IEEE ICC'97*, vol. 1, (Montreal,Canada), pp. 309–313, 1997.

[35] V. Tarokh, A. F. Naguib, N. Seshadri, and A. Calderbank, "Space-Time Codes for High Data Rate Wireless Communications: Performance Criteria in the Presence of Channel Estimation Errors, Mobility, and Multiple Paths," *IEEE Trans. Commun.*, vol. 47, pp. 199–207, February 1999.

[36] V. Tarokh, N. Seshadri, and A. R. Calderbank, "Space-Time Codes for High Data Rate Wireless Communications: Performance Criterion and Code Construction," *IEEE Trans. Inform. Theory*, pp. 744–765, March 1998.

[37] V. Tarokh, A. Naguib, N. Seshadri, and A. R. Calderbank, "Combined Array Processing and Space-Time Coding," *IEEE Trans. Inform. Theory*, pp. 1121–1128, May 1999.

[38] S. Alamouti, "Space Block Coding: A Simple Transmitter Diversity Technique for Wireless Communications," *IEEE Journal on Selec. Areas. Commun.*, vol. 16, pp. 1451–1458, October 1998.

[39] V. Tarokh, H. Jafarkhani, and R. A. Calderbank, "Space-Time Block Codes From Orthogonal Designs," *IEEE Trans. Inform. Theory*, vol. 45, pp. 1456–1467, July 1999.

[40] V. Tarokh, H. Jafarkhani, and R. A. Calderbank, "Space-Time Block Codes for High Data Rates Wireless Communications: Performance Results," *IEEE Journal on Selec. Areas. Commun.*, vol. 17, pp. 451–460, March 1999.

[41] V. Tarokh and H. Jafarkhani, "A Differential Detection Scheme for Transmit Diversity." to appear in the IEEE Journal on Selec. Areas. Commun., 2000.

[42] V. Tarokh, S. M. Alamouti, and P. Poon, "New Detection Scheme for Transmit Diversity with not Channel Estimation." to appear in the IEEE Transactions on Vehicular Technology, 2000.

[43] A. F. Naguib and N. Seshadri, "Combined Interference Cancellation and ML Decoding of Space-Time Block Codes." Accepted for Publication in IEEE Journal on Selec. Areas. Commun., 2000.

[44] A. F. Naguib, "Combined Interference Cancellation and ML Decoding of Space-Time Block Codes II: The General Case." submitted to IEEE Journal on Selec. Areas. Commun., Jan 2000.

[45] A. F. Naguib, N. Seshadri, and A. R. Calderbank, "Applications of Space-Time Block Codes and Interference Suppression for High Capacity and High Data Rate Wireless Systems," in *Proc.32nd Asilomar Conf. Signals,*

Systems, and Computers, vol. 2, (Pacific Grove, CA), pp. 1803–1810, November 1998.

[46] J. G. Proakis, *Digital Communications*. New York, NY: McGraw-Hill, second ed., 1989.

[47] A. F. Naguib, V. Tarokh, N. Seshadri, and A. R. Calderbank, "A Space-Time Coding Based Modem for High Data Rate Wireless Communications," *IEEE Journal on Selec. Areas. Commun.*, vol. 16, pp. 1459–1478, October 1998.

[48] J. K. Cavers, "An Analysis of Pilot Symbol Assisted Modulation for Rayleigh Faded Channels," *IEEE Trans. Veh. Technology*, vol. VT-40, pp. 683–693, November 1991.

[49] S. Sampei and T. Sunaga, "Rayleigh Fading Compensation Method for 16 QAM in Digital Land Mobile Radio Channels," in *Proc. IEEE VTC'89*, vol. I, (San Francisco), pp. 640–646, May 1989.

[50] M. L. Moher and J. H. Lodge, "TCMP-A Modulation and Coding Strategy for Rician Fading Channels," *IEEE Journal on Selected Areas in Commun.*, vol. JSAC-7, pp. 1347–1355, December 1989.

[51] R. J. Young, J. H. Lodge, and L. C. Pacola, "An Implementation of a Reference Symbol Approach to Generic Modulation in Fading Channels," in *Proc. of International Mobile Satellite Conf.*, (Ottawa), pp. 182–187, June 1990.

[52] J. Yang and K. Feher, "A digital Rayleigh Fade Compensation Technology for Coherent OQPSK System," in *Proc. IEEE VTC'90*, (Orlando,Fl), pp. 732–737, May 1990.

[53] C. L. Liu and K. Feher, "A New Generation of Rayleigh Fade Compensated $\frac{\pi}{4}$-QPSK Coherent Modem," in *Proc. IEEE VTC'90*, (Orlando,Fl), pp. 482–486, May 1990.

[54] A. Aghamohammadi, H. Meyr, and G. Asheid, "A New Method for Phase Synchronization and Automatic Gain Control of Linearly Modulated Signals on Frequency-Flat Fading Channel," *IEEE Trans. Commun.*, vol. COM-39, pp. 25–29, January 991.

[55] B. L. Hughes, "Differential Space-Time Modulation." Submitted to IEEE Trans. Information Theory, 2000.

[56] B. M. Hochwald and T. L. Marzetta, "Unitary Space-Time Modulation for Multiple Antenna Communications in Rayleigh Flat Fading." to appear in IEEE Trans. Information Theory, March 2000.

[57] B. M. Hochwald, T. L. Marzetta, T. J. Richardson, W. Sweldons, and R. Urbanke, "Systematic Design of Unitary Space-Time Constellation." submitted to IEEE Trans. Information Theory, 2000.

[58] J. H. Winters, J. Salz, and R. D. Gitlin, "The Impact of Antenna Diversity on the Capacity of Wireless Communication Systems," *IEEE Trans. Commun.*, vol. 42, pp. 1740–1751, February/March/April 1994.

[59] A. F. Naguib and N. Seshadri, "Combined Interference Cancellation and ML Decoding of Space-Time Block Codes," in *Communication Theory Mini Conference. Held in Conjunction with Globecomm'98*, (Sydney, Australia), pp. 7–15, November 1998.

[60] G. Raleigh and J. M. Cioffi, "Spatio-Temporal Coding for Wireless Communications," *IEEE Trans. Communications*, vol. 46, pp. 357–366, March 1998.

[61] TIA 45.5 Subcommitte, "The CDMA 2000 Candidate Submission," Draft, June 1998.

[62] R. Wichman and A. Hottinen, "Transmit Diversity WCDMA System," Technical Report, Nokia Research Center, 1998.

[63] D. Agrawal, V. Tarokh, A. Naguib, and N. Seshadri, "Space-Time Coded OFDM for High Data Rate Wireless Communication over Wideband Channels," in *Proc. of 48th IEEE VTC, 1998*, vol. 3, (Ottawa, Canada), pp. 2232–2236, May 1998.

[64] Z. Liu, G. B. Giannakis, A. Scaglione, and S. Barbarossa, "Block Precoding and Transmit-Antenna Diversity for Decoding and Equalization of Unknown Multipath Channels," in *Proc. of 33rd Asilomar Conf. on Signals, Systems, and Computers*, vol. 2, (Pacific Grove, CA), pp. 1557–1561, November 1-4 1999.

[65] Z. Liu and G. B. Giannakis, "Space-Time Coding with Transmit Antennas for Multiple Access Regardless of Frequency-Selective Multipath," in *Proc. of the 1st Sensor Array and Multichannel SP Workshop*, (Boston, MA), March 15-17 2000.

Chapter 3

CODING FOR THE WIRELESS CHANNEL

Ezio Biglieri, Giorgio Taricco

Dipartimento di Elettronica, Politecnico di Torino, Torino, Italy

ezio@polito.it, giorgio@polito.it

Giuseppe Caire

Institut Eurecom, Sophia Antipolis, France

giuseppe.caire@eurecom.fr

Abstract   In this chapter we describe some techniques for selecting coding schemes for wireless channels, and in particular the frequency-flat, slow fading channel. Optimum coding schemes for this channel lead to the development of new criteria for code design, differing markedly from the Euclidean-distance criterion which is commonplace over the additive white Gaussian noise (AWGN) channel. For example, the code performance depends strongly, rather than on the minimum Euclidean distance of the code, on its minimum Hamming distance (the "code diversity"). If the channel model is not stationary, as it happens for example in a mobile-radio communication system where it may fluctuate in time between the extremes of Rayleigh and AWGN, then a code designed to be optimum for a fixed channel model might perform poorly when the channel varies. Therefore, a code optimal for the AWGN channel may be actually suboptimum for a substantial fraction of time. In these conditions, antenna diversity with maximum-gain combining may prove useful: in fact, under fairly general conditions, a channel affected by fading can be turned into an AWGN channel by increasing the number of diversity branches. Another robust solution is based on bit interleaving, which yields a large diversity gain thanks to the choice of powerful convolutional codes coupled with a bit interleaver and the use of a suitable bit metric. An important feature of bit-interleaved coded modulation is that it lends itself quite naturally to "pragmatic" designs, i.e., to coding schemes that keep as their basic engine an off-the-shelf Viterbi decoder. Yet another solution is based on controlling the transmitted power so as to compensate for the attenuations due to fading.

Keywords: Fading channels, coding, code diversity, antenna diversity, bit-interleaved coded modulation, power control

1   INTRODUCTION

In the simplest communication channel model (the "additive white Gaussian channel", or AWGN) the received signal is assumed to be affected only by a constant attenuation and a constant delay. Digital transmission over radio channels often needs a more elaborate model, since it may be necessary to account for propagation vagaries, referred to as "fading," which affect the signal strength. These are connected with a propagation environment referred to as "multipath" and with the relative movement of transmitter and receiver, which causes time variations of the channel.

Multipath propagation occurs when the electromagnetic energy carrying the modulated signal propagates along more than one "path" connecting the transmitter to the receiver. Examples of such situation occur for example in indoor propagation when the electromagnetic waves are perturbed by structures inside the building, and in terrestrial mobile radio when multipath is caused by large fixed or moving objects (buildings, hills, cars, etc.). For an extensive review of the main aspects of wireless channels, see [1] and references therein.

2   CODING FOR THE FADING CHANNEL

Coding solutions for the fading channel should be selected by taking into account the distinctive features of the model used. Our goal here is to survey these solutions, by highlighting a number of issues that make code design for the fading channel differ from that for the AWGN channel. In this survey we examine in particular the effects of three features that make the fading channel differ from AWGN: namely, the fading channel is generally not memoryless (unless infinite-depth interleaving is assumed, an assumption that may not be realistic in several instances), has a signal-to-noise ratio which is a random variable rather than a constant, and finally the propagation vagaries may make the channel model to vary with time, so that any chosen model may be able to represent the channel only for a fraction of the time.

2.1   TURBO CODES AND THE LIKE

Discovered in 1993 by Berrou and Glavieux [2], Turbo codes have revolutionized the field of error-control codes. These codes achieve reliable communication at data rates near the ultimate capacity limits (see Fig. 3.1), and yet have enough structure to allow practical encoding and decoding algorithms.

Turbo coding consists of the combination of two key elements: constituent convolutional codes which interact in "parallel concatenation" through an interleaver, and iterative decoding. The latter is obtained by applying iteratively the BCJR algorithm [3] while isolating different information sources, thus allowing for the development of independent estimates of a posteriori distribu-

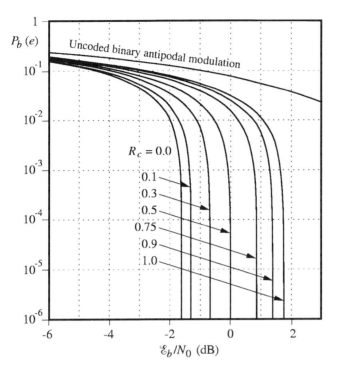

Figure 3.1 Bit error probability vs. signal-to-noise ratio over the Gaussian channel. Potential coding gains with different code rates and soft decoding.

tions of transmitted data. Of late, there has been a considerable amount of research into the relationship between turbo decoding and graphical models for codes (see, e.g, [4]). McKay and Neal [5, 6] have shown that Gallager's algorithms [7] for decoding low-density parity-check codes allow these to perform just as well as turbo codes.

The actual level of understanding of turbo codes (and of other families of codes that perform close to capacity) is still limited [8]: in particular, their performance of is mostly evaluated by simulation. The reason for this resides in an intrinsic weakness of the union bound to error probability, which is by far the most widely used tool for the prediction of code performance. This is easy to compute, and requires only the knowledge of the weight spectrum of the code; however, it becomes too loose, and hence useless, when the signal-to-noise ratio approaches the value at which the cutoff rate R_0 of the channel equals the code rate R_c. Now, for turbo codes bounds are needed that overcome the R_0-limitation of the union bound. For recent work in this area, see [9] and references therein.

2.2   SPEECH VS. DATA: THE DELAY ISSUE

A relevant factor in the choice of a coding scheme is the decoding delay that one may allow: for example, recently proposed, extremely powerful codes (the "Turbo Codes" of [2]) suffer from a considerable decoding delay, and hence their applicability is restricted.

Consider for example real-time speech transmission: here a strict decoding delay is imposed (e.g., 100 ms, at most [10]). In this case, the transmission of a code word may span only a few TDMA channel bursts, over which the channel fading is strongly correlated. Thus, a code word experiences only a few significant fading values, which makes the assumption of a memoryless channel, normally achieved by ideal or very long interleaving, no longer valid. On the contrary, with data traffic a large interleaving delay is tolerable, so that very effective coding techniques are available. For example, as we shall see, convolutional codes, bit interleaving, and high-level modulation (such as 8PSK or 16QAM) can be used. These techniques are generally referred to as Bit-Interleaved Coded Modulation (BICM) and have been extensively studied in [11] Capacity calculations show that with large interleaving BICM performs as well as optimal coding over more complicated alphabets, and its complexity is much lower, so that the performance-complexity trade-off of BICM is very attractive. Moreover, capacity calculations [12] show that constant-power constant-rate transmission performs very close to optimal transmission schemes where power and rate are adapted dynamically to the channel conditions via a perfect feedback link. Then, with large interleaving and powerful coding, there is no need for implementing such complicated adaptive techniques and feedback links.

2.3   MODELING THE DELAY CONSTRAINTS

The delay constraints can be easily taken into account when designing a coding scheme if a "block-fading" channel model is used. In this model, the fading process is about constant for a number of symbol intervals. On such a channel, a single code word may be transmitted after being split into several blocks, each suffering from a different attenuation, and thus realizing an effective way of achieving diversity.

The "block-fading" channel model, introduced in [10, 13], is motivated by the fact that, in many mobile radio situations, the channel coherence time is much longer than one symbol interval, and hence several transmitted symbols are affected by the same fading value. Use of this channel model allows one to introduce a delay constraint for transmission, which is realistic whenever infinite-depth interleavinginterleaver is not a reasonable assumption.

This model assumes that a code word of length $n = MN$ spans M blocks of length N (a group of M blocks will be referred to as a *frame*.) The value of the fading in each block is constant. M turns out to be a measure of the interleaving *delay* of the system: in fact, $M = 1$ corresponds to $N = n$, i.e., to no interleaving, while $M = n$ corresponds to $N = 1$, and hence to ideal interleaving. Thus, the results for different values of M illustrate the downside of nonideal interleaving. It should also be observed that the coding scheme implied by this channel model generalizes standard diversity techniques: in fact, the latter can be seen as a special case of coding for a block-fading channel on which repetition codes are used.

With no delay constraint, a code word can span an arbitrarily large number M of fading blocks. If this is the case, then capacity, as derived in [12], is a good performance indicator. This applies for example to variable-rate systems (e.g., wireless data networks). On the other hand, most of today's mobile radio systems carry real-time speech (cellular telephony), for which constant-rate, constrained-delay transmission should be considered. In the latter case, that is, when each code word must be transmitted and decoded within a frame of $M < \infty$ blocks, *information outage rate*, rather than capacity, is the appropriate performance limit indicator. We shall not delve in this issue any further here, and the interested reader is referred to [10, 14].

2.4   DIVERSITY

Receiver-diversity techniques have been known for a long time to improve the fading-channel quality. Recently, their synergy with coding has been extensively investigated in [15, 16, 17]. The standard approach to antenna diversity is based on the fact that, with several diversity branches, the probability that the signal will be simultaneously faded on all branches can be made small. The approach taken in [15, 16, 17, 18] is philosophically different, as it is based upon the observation that, under fairly general conditions, a channel affected by fading can be turned into an additive white Gaussian noise (AWGN) channel by increasing the number of diversity branches. Consequently, it can be expected (and it was indeed verified by analyses and simulations) that a coded modulation scheme designed to be optimal for the AWGN channel will perform asymptotically well also on a fading channel with diversity, at the only cost of an increased receiver complexity. An advantage of this solution is its robustness, since changes in the physical channel affect the reception very little.

This allows us to argue that the use of "Gaussian" codes along with diversity reception provides indeed a solution to the problem of designing robust coding schemes for the mobile radio channel.

2.5   UNEQUAL ERROR PROTECTION

In some analog source coding applications, like speech or video compression, the sensitivity of the source decoder to errors in the coded symbols is typically not uniform: the quality of the reconstructed analog signal is rather insensitive to errors affecting certain classes of bits, while it degrades sharply when errors affect other classes. This happens, for example, when analog source coding is based on some form of hierarchical coding, where a relatively small number of bits carries the "fundamental information" and a larger number of bits carries the "details" like in the case of the MPEG2 standard.

Assuming that the source encoder produces frames of binary coded symbols, each frame can be partitioned into classes of symbols of different "importance" (i.e., of different sensitivity). Then, it is apparent that the best coding strategy aims at achieving lower BER levels for the important classes while admitting higher BER levels for the unimportant ones. This feature is referred to as unequal error protection (UEP). On the contrary, codes for which the BER is (almost) independent of the position of the information symbols are referred to as equal error protection (EEP) codes.

An efficient method for achieving UEP with Turbo Codes was recently studied in [19]. The key point is to match a non-uniform puncturing pattern to the interleaver of the Turbo-encoder in order to create locally low-rate Turbo Codes for the important symbols, and locally high-rate Turbo Codes for the unimportant symbols. In this way, we can achieve several protection levels while keeping constant the total code rate. On the decoding side, all what we need is to "depuncture" the received sequence by inserting zeros at the punctured positions. Then, a single Turbo-decoder can handle different code rates, equal-error-protection Turbo Codes and UEP Turbo Codes.

2.6   THE FREQUENCY-FLAT, SLOW RAYLEIGH-FADING CHANNEL

This channel model assumes that the duration of a modulated symbol is much greater than the delay spread caused by the multipath propagation. If this occurs, then all frequency components in the transmitted signal are affected by the same random attenuation and phase shift, and the channel is frequency-flat. If in addition the channel varies very slowly with respect the symbol duration, then the fading $R(t)\exp[j\Theta(t)]$ remains approximately constant during the transmission of one symbol (if this does not occur the fading process is called *fast*.)

The assumption of non-selectivity allows us to model the fading as a process affecting the transmitted signal in a multiplicative form. The assumption of a slow fading allows us to model this process as a constant random variable during each symbol interval. In conclusion, if $x(t)$ denotes the complex envelope of the modulated signal transmitted during the interval $(0, T)$, then the complex envelope of the signal received at the output of a channel affected by slow, flat fading and additive white Gaussian noise can be expressed in the form

$$r(t) = Re^{j\Theta}x(t) + n(t), \qquad (3.1)$$

where $n(t)$ is a complex Gaussian noise, and $Re^{j\Theta}$ is a Gaussian random variable, with R having a Rice or Rayleigh pdf and unit second moment, i.e., $E[R^2] = 1$.

If we can further assume that the fading is so slow that we can estimate the phase shift Θ with sufficient accuracy, and hence compensate for it, then coherent detection is feasible. Thus, model (3.1) can be further simplified to

$$r(t) = Rx(t) + n(t). \qquad (3.2)$$

It should be immediately apparent that with this simple model of fading channel the only difference with respect to an AWGN channel resides in the fact that R, instead of being a constant attenuation, is now a random variable, whose value affects the amplitude, and hence the power, of the received signal. Assume finally that the value taken by R is known at the receiver: we describe this situation by saying that we have *perfect* CSI. Channel state information can be obtained for example by inserting a pilot tone in a notch of the spectrum of the transmitted signal, and by assuming that the signal is faded exactly in the same way as this tone.

Detection with perfect CSI can be performed exactly in the same way as for the AWGN channel: in fact, the constellation shape is perfectly known, as is the attenuation incurred by the signal. The optimum decision rule in this case consists of minimizing the Euclidean distance

$$\int_0^T [r(t) - Rx(t)]^2 dt \quad \text{or} \quad |\mathbf{r} - R\mathbf{x}|^2 \qquad (3.3)$$

with respect to the possible transmitted real signals $x(t)$ (or vectors $\mathbf{x}$).

A consequence of this fact is that the error probability with perfect CSI and coherent demodulation of signals affected by frequency-flat, slow fading can be evaluated as follows. We first compute the error probability $P(e \mid R)$ obtained by assuming R constant in model (3.2), then we take the expectation of $P(e \mid R)$, with respect to the random variable R. The calculation of $P(e \mid R)$ is performed as if the channel were AWGN, but with the energy $\mathcal{E}$ changed into $R^2\mathcal{E}$. Notice finally that the assumptions of a noiseless channel-state information and a noiseless phase-shift estimate make the values of $P(e)$ thus obtained as representing a limiting performance.

Consider now the error probabilities that we would obtain with binary signals without coding (see [20] for a more general treatment). For example, for

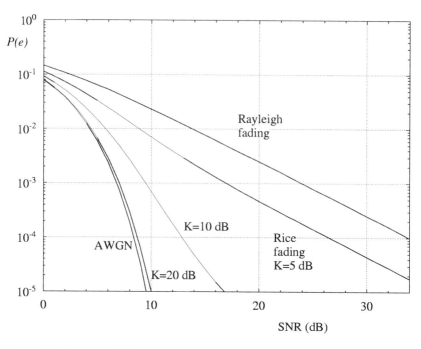

Figure 3.2   Error probabilities of binary antipodal transmission over the Gaussian channel and over Rayleigh and Rice fading channels.

two antipodal signals with common energy $\mathcal{E}$ we have, for Rayleigh fading and perfect channel-state information,

$$P(e) = \frac{1}{2}\left(1 - \sqrt{\frac{\mathcal{E}/N_0}{1 + \mathcal{E}/N_0}}\right). \tag{3.4}$$

In the absence of CSI, one could take a decision rule consisting of minimizing

$$\int_0^T [r(t) - x(t)]^2 \, dt \quad \text{or} \quad |\mathbf{r} - \mathbf{x}|^2 \tag{3.5}$$

However, with constant envelope signals ($|\mathbf{x}|$ constant), the error probability obtained with (3.3) and (3.5) coincide because

$$
\begin{aligned}
P(\mathbf{x} \to \widehat{\mathbf{x}}) &= P(|\mathbf{r} - R\widehat{\mathbf{x}}|^2 < |\mathbf{r} - R\mathbf{x}|^2) \\
&= P(2R(\mathbf{r}, \mathbf{x} - \widehat{\mathbf{x}}) < 0) \\
&= P((\mathbf{r}, \mathbf{x} - \widehat{\mathbf{x}}) < 0)
\end{aligned}
$$

and hence CSI is completely represented by the phase Θ. Fig. 3.2 compares error probabilities of binary antipodal transmission over the Gaussian channel with those over the Rayleigh and Rice fading channel, where K denotes the "Rice factor" of the latter [21]. Although the loss in error probability is considerable, coding can compensate for a substantial fraction of this loss.

2.7   OUR SURVEY

In this chapter we survey a few important issues in coding for the fading channel. The model we assume here is that of a channel affected by flat, slow fading and additive noise. Optimum coding schemes for this channel model lead to the development of new criteria for code design (Section 3). If the channel model is not stationary, as it happens for example in a mobile-radio communication system, then a code designed to be optimum for a fixed channel model might perform poorly when the channel varies. Therefore, a code optimal for the AWGN channel may be actually suboptimum for a substantial fraction of time. In these conditions, antenna diversity with maximum-gain combining may prove useful: in fact, under fairly general conditions, a channel affected by fading can be turned into an AWGN channel by increasing the number of diversity branches (Section 4.1). Another robust solution is based on bit interleaving, which yields a large diversity gain thanks to the choice of powerful convolutional codes coupled with a bit interleaver and the use of a suitable bit metric (Section 4.2). Yet another solution is based on controlling the transmitted power so as to compensate for the attenuations due to fading (Section 4.3).

3   CODE-DESIGN CRITERIA

A standard code-design criterion, when soft decoding is chosen, is to choose coding schemes that maximize their minimum Euclidean distance. This is of course correct on the Gaussian channel with high SNR (although not when the SNR is very low), and is often accepted, *faute de mieux*, on channels that deviate little from the Gaussian model (e.g., channels with a moderate amount of intersymbol interference). However, the Euclidean-distance criterion should be outright rejected over the Rayleigh fading channel. In fact, analysis of coding for the Rayleigh fading channel proves that Hamming distance (also called "code diversity" in this context) plays the central role here.

Assume transmission of a coded sequence $X = (x_1, x_2, \ldots, x_n)$ where the components of X are signal vectors selected from a constellation S. We do not distinguish here among block or convolutional codes (with soft decoding), or block- or trellis-coded modulation.

3.1   NO DELAY CONSTRAINT: INFINITE-DEPTH INTERLEAVING

We also assume for the moment infinite-depth interleaving, which makes the fading random variables affecting the various symbols x_k to be independent. Hence we write, for the components of the received sequence $(r_1, r_2, \dots, r_n)$:

$$r_k = R_k x_k + n_k, \tag{3.6}$$

where the R_k are independent, and, under the assumption that the noise is white, the RV's n_k are also independent.

Coherent detection of the coded sequence, with the assumption of perfect channel-state information, is based upon the search for the coded sequence X that minimizes the distance

$$\sum_{k=1}^{N} |r_k - R_k x_k|^2. \tag{3.7}$$

The pairwise error probability can be upper bounded in this case as [21, Chap.13], [22]

$$P\{X \to \widehat{X}\} \le \prod_{k \in \mathcal{K}} \frac{1}{1 + |x_k - \widehat{x}_k|^2/4N_0} \tag{3.8}$$

where $\mathcal{K}$ is the set of indices k such that $x_k \ne \widehat{x}_k$.

An example. For illustration purposes, let us compute the Chernoff upper bound to the word error probability of a block code with rate R_c. Assume that binary antipodal modulation is used, with waveforms of energies $\mathcal{E}$, and that the demodulation is coherent with perfect CSI. Observe that for $\widehat{x}_k \ne x_k$ we have

$$|x_k - \widehat{x}_k|^2 = 4\mathcal{E} = 4R_c\mathcal{E}_b,$$

where $\mathcal{E}_b$ denotes the average energy per bit. For two code words X, $\widehat{X}$ at Hamming distance $d_H(X, \widehat{X})$ we have

$$P\{X \to \widehat{X}\} \le \left(\frac{1}{1 + R_c\mathcal{E}_b/N_0}\right)^{d_H(X, \widehat{X})}$$

and hence, for a linear code,

$$P(e) = P(e \mid X) \le \sum_{w \in \mathcal{W}} \left(\frac{1}{1 + R_c\mathcal{E}_b/N_0}\right)^{w},$$

where $\mathcal{W}$ denotes the set of nonzero Hamming weights of the code, considered with their multiplicities. It can be seen that for high enough signal-to-noise ratio the dominant term in the expression of $P(e)$ is the one with exponent $d_{\min}$, the minimum Hamming distance of the code. □

By recalling the above calculation, the fact that the probability of error decreases inversely with the signal-to-noise ratio raised to power $d_{\min}$ can be expressed by saying that we have introduced a *code diversity* $d_{\min}$.

We may further upper bound the pairwise error probability by defining the set $\mathcal{K}$ of indices k for which $x_k \neq \hat{x}_k$, and writing

$$P\{X \to \hat{X}\} \le \prod_{k \in \mathcal{K}} \frac{1}{|x_k - \hat{x}_k|^2 / 4N_0} = \frac{1}{[\delta^2(X, \hat{X})/4N_0]^{d_H(X, \hat{X})}} \qquad (3.9)$$

(which is close to the true Chernoff bound for small enough N_0). Here

$$\delta^2(X, \hat{X}) = \left[\prod_{k \in \mathcal{K}} |x_k - \hat{x}_k|^2 \right]^{1/d_H(X, \hat{X})}$$

is the geometric mean of the non-zero squared Euclidean distances between the components of X, $\hat{X}$. The latter result shows the important fact that the error probability is (approximately) inversely proportional to the *product* of the squared Euclidean distances between the components of $\mathbf{x}$, $\hat{\mathbf{x}}$ that differ, and, to a more relevant extent, to a power of the signal-to-noise ratio whose exponent is the Hamming distance between X and $\hat{X}$.

Further, we know from the results referring to block codes, convolutional codes, and coded modulation that the union bound to error probability for a coded system can be obtained by summing up the pairwise error probabilities associated with all the different "error events." For small noise spectral density N_0, i.e., for high signal-to-noise ratios, a few equal terms will dominate the union bound. These correspond to error events with the smallest value of the Hamming distance $d_H(X, \hat{X})$. We denote this quantity by L_c to stress the fact, to be discussed soon, that it reflects a diversity residing in the code. We have

$$P\{X \to \hat{X}\} \lesssim \frac{\nu}{[\delta^2(X, \hat{X})/4N_0]^{L_c}} \qquad (3.10)$$

where ν is the number of dominant error events. For error events with the same Hamming distance, the values taken by $\delta^2(X, \hat{X})$ and by ν are also of importance. This observation may be used to design coding schemes for the Rayleigh fading channel: here no role is played by the Euclidean distance, which is the central parameter used in the design of coding schemes for the AWGN channel.

For uncoded systems ($n = 1$), the results above hold with the positions $L_c = 1$ and $\delta^2(X, \hat{X}) = |\mathbf{x} - \hat{\mathbf{x}}|^2$, which shows that the error probability decreases as N_0. A similar result could be obtained for maximal-ratio combining in a system with diversity L_c. This explains the name of this parameter. In this context, the various diversity schemes may be seen as implementations of the simplest among the coding schemes, the repetition code, which provides a diversity equal to the number of diversity branches [22].

From the discussion above, we have learned that over the perfectly-interleaved Rayleigh fading channel the choice of a coding scheme should be based on the maximization of the code diversity, i.e., the minimum Hamming distance among pairs of error events. Since for the Gaussian channel code diversity does not play the same central role, coding schemes optimized for the Gaussian channel are likely to be suboptimum for the Rayleigh channel.

3.2   INTRODUCING DELAY CONSTRAINTS: THE BLOCK-FADING CHANNEL

The above analysis holds, *mutatis mutandis*, for the block-fading channel: it suffices in this case to interpret the variables x_k as *blocks of symbols*, rather than symbols. In this situation, it should not come as a surprise (and can in fact be shown rigorously, see [23, 24]) that the relevant criterion becomes the *block-Hamming* distance, i.e., the number of *blocks* in which two code words differ. An application of Singleton Bound shows that the maximum block-Hamming distance achievable on an M-block fading channel is limited by

$$D \leq 1 + \left\lfloor M \left(1 - \frac{R}{\log_2 |\mathcal{S}|} \right) \right\rfloor$$

where $|\mathcal{S}|$ is the size of the signal set $\mathcal{S}$ and R is the code rate, expressed in bit/symbol. Note that binary signal sets ($|\mathcal{S}| = 2$) are not effective in this case, so that codes constructed over high-level alphabets should be considered [23, 24].

For a deeper analysis of the relationship between code diversity and code rate, see [25, 26].

4   ROBUST CODING SCHEMES

The design procedure described in the section above, and consisting of adapting the coding scheme to the channel, may suffer from a basic weakness. If the channel model is not stationary, as it is, for example, in a mobile-radio environment where it fluctuates in time between the extremes of Rayleigh and AWGN, then a code designed to be optimum for a fixed channel model might perform poorly when the channel varies. Therefore, a code optimal for the AWGN channel may be actually suboptimum for a substantial fraction of time. An alternative solution consists of doing the opposite, i.e., *matching the channel to the coding scheme*: the latter is still designed for a Gaussian channel, while the former is transformed from a Rayleigh-fading channel (say) into a

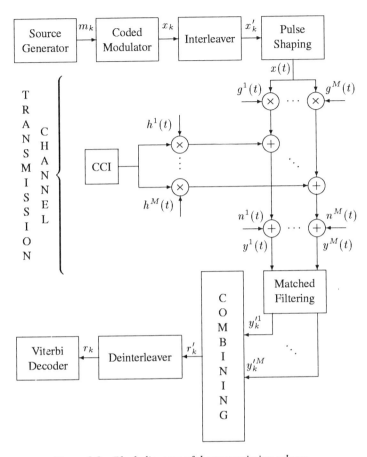

Figure 3.3   Block diagram of the transmission scheme.

Gaussian one. Here we shall examine three such robust solutions, the first based on antenna diversity, the second on bit-interleaving, and the third on power control.

4.1   ANTENNA DIVERSITY

Fig. 3.3 shows the block diagram of the transmission scheme with fading. A source of co-channel interference is also added for completeness. Our initial assumptions, valid in the following unless otherwise stated, are [15, 16, 17]:

- PSK modulation

- M independent diversity branches whose signal-to-noise ratio is inversely proportional to M (this assumption is made in order to disregard the

SNR increase that actually occurs when multiple receive elements are used).

- Flat, independent Rayleigh fading channel.

- Coherent detection with perfect channel-state information.

- Synchronous diversity branches.

- Independent co-channel interference, and a single interferer.

The codes examined are the following:

J4: 4-state, rate-2/3 TCM scheme based on 8-PSK and optimized for Rayleigh fading channels [27].

U4: 4-state rate-2/3 TCM scheme based on 8-PSK and optimized for the Gaussian channel.

U8: Same as above, with 8 states.

Q64: "Pragmatic" concatenation of the "best" rate-1/2 64-state convolutional code with 4-PSK modulator and Gray mapping [28].

Fig. 3.4 compares the performance of U4 and J4 (two TCM schemes with the same complexity) over a Rayleigh-fading channel with M-branch diversity. It is seen that, as M increases, the performance of U4 comes closer and closer to that of J4. Similar results hold for correlated fading: even for moderate correlation, J4 loses its edge on U4, and for M as low as 4, U4 performs better than J4 [15]. The effect of diversity is more marked when the code used is weaker. As an example, two-antenna diversity provides a gain of 10 dB at BER=10^{-6} when U8 is used, and of 2.5 dB when Q64 is used [15]. The assumption of branch independence, although important, is not critical: in effect, [15] shows that branch correlation coefficients as large as 0.5 degrade system BER only slightly. The complexity introduced by diversity can be traded for delay: as shown in [15], in some cases diversity makes interleaving less necessary, so that a lower interleaving depth (and consequently a lower overall delay) can be compensated by an increase of M.

When differential or pilot-tone, rather than coherent, detection is used [16], a BER-floor occurs which can be reduced by introducing diversity. As for the effect of co-channel interference, even its BER-floor is reduced as M increased (although for its elimination multi-user detectors should be employed). This shows that antenna diversity with maximal-ratio combining is highly instrumental in making the fading channel closer to Gaussian.

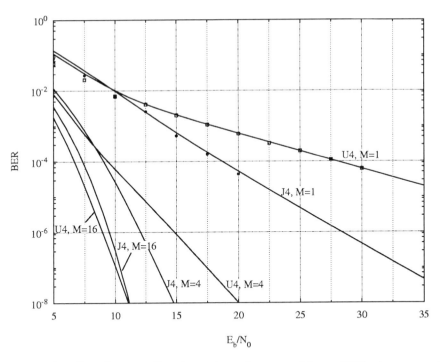

Figure 3.4   Effect of antenna diversity on the performance of 4-state TCM schemes over the flat, independent Rayleigh-fading channel. J4 is optimum for the Rayleigh channel, while U4 is optimum for the Gaussian channel.

4.2   BIT-INTERLEAVED CODED MODULATION

Ever since 1982, when Ungerboeck published his landmark paper on trellis-coded modulation [29], it has been generally accepted that modulation and coding should be combined in a single entity for improved performance. Several results followed this line of thought, as documented by a considerable body of work aptly summarized and referenced in [27] (see also [30, Chap. 10]). Under the assumption that the symbols were interleaved with a depth exceeding the coherence time of the fading process, new codes were designed for the fading channel so as to maximize their diversity. This implied in particular that parallel transitions should be avoided in the code, and that any increase in diversity would be obtained by increasing the constraint length of the code. One should also observe that for non-Ungerboeck systems, i.e., those separating modulation and coding with binary modulation, Hamming distance is proportional to Euclidean distance, and hence a system optimized for the additive white Gaussian channel is also optimum for the Rayleigh fading channel.

A notable departure from Ungerboeck's paradigm was the core of [28]. Schemes were designed in which coded modulation is generated by pairing an M-ary signal set with a binary convolutional code with the largest minimum free Hamming distance. Decoding was achieved by designing a metric aimed at keeping as their basic engine an off-the-shelf Viterbi decoder for the *de facto* standard, 64-state rate-$1/2$ convolutional code. This implied giving up the joint decoder/demodulator in favor of two separate entities.

Based on the latter concept, Zehavi [31] first recognized that the code diversity, and hence the reliability of coded modulation over a Rayleigh fading channel, could be further improved. Zehavi's idea was to make the code diversity equal to the smallest number of distinct *bits* (rather than *channel symbols*) along any error event. This is achieved by bit-wise interleaving at the encoder output, and by using an appropriate soft-decision bit metric as an input to the Viterbi decoder. For different approaches to the problem of designing coded modulation schemes for the fading channels see [32].

One of Zehavi's findings, rather surprising *a priori*, was that on some channels, there is a downside to combining demodulation and decoding. This prompted the investigation whose results are presented in a comprehensive fashion in [11].

An advantage of this solution is its robustness, since changes in the physical channel affect the reception very little. Thus, it provides good performance with a fading channel as well as with an AWGN channel (and, consequently, with a Rice fading channel, which can be seen as intermediate between the latter two).

4.3   POWER CONTROL

Observation of (3.2) shows that what makes this Rayleigh fading channel differ from AWGN is the fact that R is a random variable, rather than a constant attenuation. Consequently, if this variability of R could be compensated for, an AWGN would be obtained. This compensation can be achieved in principle if channel-state information is available to the transmitter, which consequently can modulate its power according to the channel fluctuations.

Consider the simplest such strategy. The flat, independent fading channel with coherent detection yields the received signal (3.2). Assume that the channel state information R is known at the transmitter front-end, that is, the transmitter knows the value of R during the transmission (this assumption obviously requires that R is changing very slowly). Under these conditions, assume that the transmitted signal in an interval with length T is

$$x(t) = \sigma s(t), \tag{3.11}$$

where $s(t)$ has unit energy (equal-energy basic waveform), and σ is chosen under a given optimality criterion.

One possible such criterion (constant error probability over each symbol) requires that

$$\sigma = R^{-1}. \tag{3.12}$$

This way, the channel is transformed into an equivalent additive white Gaussian noise channel. The error probability is the same as if we had transmitted s over a channel whose only effect is the addition of n to the transmitted signal. The average transmitted power per symbol is then

$$\mathrm{E}[x^2(t)] = \mathrm{E}[1/\rho^2], \tag{3.13}$$

which might diverge.

This technique ("channel inversion") is simple to implement, since the encoder and decoder are designed for the AWGN channel, independent of the fading statistics: for instance, it is common in spread-spectrum systems with near-far interference imbalances. However, it may suffer from a large capacity penalty. For example, in Rayleigh fading the capacity is zero.

To avoid divergence of the average power (or an inordinately large value thereof) a possible strategy is the following. Choose

$$\sigma = \begin{cases} R^{-1} & \text{if } R > R_0 \\ R_0^{-1} & \text{otherwise.} \end{cases} \tag{3.14}$$

By choosing appropriately the value of the threshold R_0 we trade off a decrease of the average power value for an increase of error probability. The average power value is now

$$(1 - p)\frac{1}{R_0^2} + p\,\mathrm{E}[1/R^2 \mid R > R_0], \tag{3.15}$$

where $p = \mathrm{P}[R > R_0]$. For an information-theoretical analysis of power-control techniques for the fading channel, see [33].

5   CONCLUSIONS

This chapter was aimed at illustrating some concepts that make the design of codes for the fading channel differ markedly from the same task applied to the Gaussian channel. In particular, we have examined the design of "fading codes," i.e., coding schemes which maximize the Hamming, rather than the Euclidean, distance, the interaction of antenna diversity with coding (which makes the channel more Gaussian), the effect of separating coding from modulation in favor of a more robust coding scheme, and the effect of transmitter-power control. The issue of optimality as contrasted to robustness was also discussed to some extent.

References

[1] E. Biglieri, J. Proakis, and S. Shamai (Shitz), "Fading channels: Information-theoretic and communications aspects," *IEEE Trans. Inform. Theory*, Vol. 44, No. 6, pp. 2619–2692, October 1998.

[2] C. Berrou and A. Glavieux, "Near optimum error correcting coding and decoding: Turbo-codes," *IEEE Trans. Commun.*, Vol. 44, No. 10, pp. 1261–1271, October 1996.

[3] L. R. Bahl, J. Cocke, F. Jelinek, and J. Raviv, "Optimal decoding of linear codes for minimizing symbol error rate," *IEEE Trans. Inform. Theory*, Vol. 20, pp. 284–287, March 1974.

[4] N. Wiberg, H.-A. Loeliger, and R. Kotter, "Codes and iterative decoding on general graphs," *European Transactions on telecommunications*, Vol. 6, No. 5, pp. 513–525, September–October 1995.

[5] D. J. C. MacKay and R. M. Neal, "Near Shannon limit performance of low density parity check codes," *Electron. Lett.*, Vol. 32, No. 18, pp. 1645–1646. Reprinted in *Electron. Lett.*, Vol. 33, pp. 457–458, Mar. 1997.

[6] D. J. C. MacKay, "Good error-correcting codes based on very sparse matrices," *IEEE TRans. Inform. Theory*, Vol. 45, No. 2, pp. 399–431, March 1999.

[7] R. G. Gallager, *Low-Density Parity-Check Codes*. Cambridge, MA: MIT Press, 1963.

[8] A. Viterbi, "Approaching the Shannon limit: Theorist's dream and practitioner's challenge," in: F. Vatalaro and F. Ananasso (Eds.), *Mobile and Personal Satellite Communications 2*, pp. 1–11. London: Springer, 1996.

[9] D. Divsalar and E. Biglieri, "Upper bounds to error probabilities of coded systems beyond the cutoff rate," *paper in preparation*.

[10] L. Ozarow, S. Shamai, and A. D. Wyner, "Information theoretic considerations for cellular mobile radio," *IEEE Trans. Vehic. Tech.*, Vol. 43, No. 2, May 1994.

[11] G. Caire, G. Taricco, and E. Biglieri, "Bit-interleaved coded modulation," *IEEE Trans. Inform. Theory*, May 1998.

[12] A. Goldsmith and P. Varaiya, "Capacity of fading channels with channel side information," *IEEE Trans. Inform. Theory*, to appear, 1997.

[13] G. Kaplan and S. Shamai (Shitz), "Error probabilities for the block-fading Gaussian channel," *A.E.U.*, Vol. 49, No. 4, pp. 192 – 205, 1995.

[14] D. Falconer and G. Foschini, "Theory of minimum mean-square-error QAM systems employing decision feedback equalization," *Bell System Tech. J.* 52, pp. 1821 – 1849, 1973.

[15] J. Ventura-Traveset, G. Caire, E. Biglieri, and G. Taricco, "Impact of diversity reception on fading channels with coded modulation—Part I: Coherent detection," *IEEE Trans. Commun.*, Vol. 45, No. 5, May 1997.

[16] J. Ventura-Traveset, G. Caire, E. Biglieri, and G. Taricco, "Impact of diversity reception on fading channels with coded modulation—Part II: Differential block detection," *IEEE Trans. Commun.*, Vol. 45, No. 6, June 1997.

[17] J. Ventura-Traveset, G. Caire, E. Biglieri, and G. Taricco, "Impact of diversity reception on fading channels with coded modulation—Part III: Co-channel interference," *IEEE Trans. Commun.*, Vol. 45, No. 7, July 1997.

[18] G. Taricco, E. Biglieri, and G. Caire, "Impact of channel-state information on coded transmission over fading channels with diversity reception," *IEEE Trans. Commun.*, Vol. 47, No. 9, pp. 1284–1287, September 1999.

[19] G. Caire and G. Lechner, "Turbo-codes with unequal error protection," *IEE Electronics Letters*, Vol. 32 No. 7, pp. 629, March 1996.

[20] E. Biglieri, G. Caire, and G. Taricco, "Error probability over fading channels: A unified approach," *European Transactions on Telecommunications*, January 1998.

[21] S. Benedetto and E. Biglieri, *Principles of Digital Transmission with Wireless Applications*. New York: Kluwer/Plenum, 1999.

[22] N. Seshadri and C.-E. W. Sundberg, "Coded modulations for fading channels—An overview," *European Trans. Telecomm.*, Vol. ET-4, No. 3, pp. 309–324, May-June 1993.

[23] R. Knopp, *Coding and Multiple Access over Fading Channels*, Ph.D. Thesis, Ecole Polytechnique Federale de Lausanne, Lausanne, Switzerland, 1997.

[24] R. Knopp, P. A. Humblet, "Maximizing diversity on block fading channels," *Proceedings of ICC '97*, Montreal, Canada, 8 – 12 June, 1997.

[25] E. Malkamaki and H. Leib, "Coded diversity schemes on block fading Rayleigh channels," *IEEE Int. Conf. on Universal Personal Communications, ICUPC'97*, San Diego, CA, October 1997.

[26] E. Malkamaki and H. Leib, "Coded diversity on block fading channels," *Submitted to IEEE Trans. Inform. Theory*, 1997.

[27] S. H. Jamali and T. Le-Ngoc, *Coded-Modulation Techniques for Fading Channels*. New York: Kluwer Academic Publishers, 1994.

[28] A. J. Viterbi, J. K. Wolf, E. Zehavi, and R. Padovani, "A pragmatic approach to trellis-coded modulation," *IEEE Communications Magazine*, vol. 27, n. 7, pp. 11–19, 1989.

[29] G. Ungerboeck, "Channel coding with multilevel/phase signals", *IEEE Trans. Inform. Th.*, vol. IT-28, pp. 56–67, Jan. 1982.

[30] E. Biglieri, D. Divsalar, P. J. McLane, and M. K. Simon, *Introduction to Trellis-Coded Modulation with Applications*, New York: Macmillan, 1991.

[31] E. Zehavi, "8-PSK Trellis Codes for a Rayleigh channel," *IEEE Trans. Commun.*, vol. 40, no. 5, pp. 873-884, May 1992.

[32] J. Boutros, E. Viterbo, C. Rastello, and J.-C. Belfiore, "Good lattice constellations for both Rayleigh fading and Gaussian channels," *IEEE Trans. Inform. Theory*, vol. 42, no. 2, pp. 502–518, March 1996.

[33] G. Caire, G. Taricco, and E. Biglieri, "Optimum power control over fading channels," *IEEE Trans. Inform. Theory*, Vol. 45, No. 5, pp. 1468–1489, July 1999.

[34] H. Asakura and T. Matsumoto, "Cooperative signal reception and downlink beamforming in cellular mobile communications," *Proceeding of ISIT'97*, Ulm, Germany, 1997.

[35] M. Fitz, J. Grimm and J. Krogmeier, "Results on code design for transmitter diversity in fading," *Proceedings of ISIT'97*, Ulm, Germany, 1997.

[36] R. Knopp and G. Caire, "Simple power-controlled modulation schemes for fixed rate transmission over fading channels," to be submitted to *IEEE Trans. Commun.*, 1997.

[37] A. Narula, M. D. Trott and G. Wornell, "Information theoretic analysis of multiple antenna transmitter diversity," submitted to *IEEE Trans. Inform. Theory*, Nov. 1996.

Chapter 4

OFDM

The Most Elegant Solution for Wireless Digital Transmission

Shinsuke Hara

Department of Electronics, Information Systems and Energy Engineering,
Graduate School of Engineering, Osaka University, Japan
hara@comm.eng.osaka-u.ac.jp

Abstract   Orthogonal Frequency Division Multiplexing (OFDM) has become part of standards in various fields such as data transmission over telephone line, digital audio
broadcasting, mobile communications and so on.

This chapter presents the principle of the OFDM, and discusses several synchronization techniques essential for successful digital transmission in multipath
fading channels, such as frequency offset estimation/compensation, window timing estimation/recovery and subcarrier recovery.

Keywords: Multicarrier transmission, frequency offset compensation, subcarrier recovery.

1   INTRODUCTION

Multi-Carrier Modulation (MCM) is the principle of transmitting data by dividing the stream into several bit streams, each of which has a much lower bit rate, and by using these substreams to modulate several subcarriers. The first systems using MCM were military HF radio links in the late 1950's and early 1960's [1]. A special form of MCM, Orthogonal Frequency Division Multiplexing (OFDM), with densely spaced subcarriers, was patented in the U.S. in 1970 [2]. OFDM abandoned the use of steep bandpass filters that completely separated the spectrum of individual subcarriers, instead, its time-domain waveforms are chosen such that mutual orthogonality is ensured even though spectra overlap. It is because such a waveform can be easily generated using the Discrete Fourier Transform (DFT) at the transmitter and receiver. However, for a relatively long time, the practicality of the concept appeared limited. Implementation aspects such as the complexity of the DFT appeared prohibitive, not to speak about other problems, such as the stability of oscillators in the transmitter and receiver, the linearity required in RF power amplifiers, and the power back-off associated with this. After many years of further intensive research in the 1980's for mobile communications [3], digital audio broadcasting [4, 5], data transmission over telephone line [6] and so on, today we appear to be on the verge of a breakthrough in OFDM techniques. Many of the implementational problems appear solvable and OFDM has become part of several standards.

This chapter focuses attention on its synchronization issue. In section 2, the basic principle of OFDM is presented, and the optimum number of subcarriers and the optimum length of guard interval are discussed. In section 3, pilot-assisted synchronization approach is presented, such as frequency offset estimation/compensation frequency domain pilot assisted subcarrier recovery and time domain pilot assisted subcarrier recovery methods, whereas in section 4, blind synchronization approach is discussed, such as joint frequency offset/symbol timing/symbol period estimation and subcarrier recovery methods.

2   PRINCIPLE AND DESIGN OF OFDM SYSTEM

Figs. 4.1 (a) and (b) show the basic block diagrams of an OFDM transmitter and receiver with N_{SC} subcarriers, respectively.

In the transmitter, the transmitted signal $s(t)$ is written as

$$s(t) = \sum_{i=-\infty}^{\infty} \sum_{k=1}^{N_{SC}} \Re[c_{ki} e^{j2\pi f_k(t-iT_s)}] f(t-iT_s). \qquad (4.1)$$

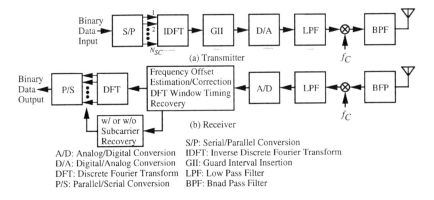

(a) Transmitter

(b) Receiver

S/P: Serial/Parallel Conversion
A/D: Analog/Digital Conversion   IDFT: Inverse Discrete Fourier Transform
D/A: Digital/Analog Conversion   GII: Guard Interval Insertion
DFT: Discrete Fourier Transform   LPF: Low Pass Filter
P/S: Parallel/Serial Conversion   BPF: Bnad Pass Filter

Figure 4.1   Basic block diagram of OFDM system.

In Eq.(4.1), $\Re[\cdot]$ is the real part of $\cdot$ ($\Im[\cdot]$ the imaginary part of $\cdot$), and f_k is the frequency of the k–th subcarrier:

$$f_k = f_C + \frac{k}{t_s}, \tag{4.2}$$

where f_C is the carrier frequency. $f(t)$ is the pulse waveform of each symbol defined as

$$f(t) = \begin{cases} 1 & (-\Delta \le t \le t_s) \\ 0 & (t < -\Delta, t > t_s), \end{cases} \tag{4.3}$$

where Δ and t_s are the guard interval and the useful symbol duration, respectively, $T_s = \Delta + t_s$ is the total symbol duration, and c_{ki} is an output at the k–th subcarrier in time $[iT_s - \Delta, iT_s + t_s]$.

The transmitted signal $s(t)$ is the sum of N_{SC} M–ary DPSK or M–ary QAM signals, and the required bandwidth is given by

$$B = \frac{2}{T_s} + \frac{N_{SC} - 1}{t_s} \approx \frac{N_{SC} + 1}{T_s} = \frac{N_{SC} + 1}{N_{SC}} R, \tag{4.4}$$

where R $(=N_{SC}/T_s)$ is the total transmission rate ($symbols/sec$).

As the number of subcarriers (N_{SC}) increases, the transmission performance becomes more sensitive to time-selectivity because the wider symbol duration is less robust to random FM noise. On the other hand, as N decreases, it becomes poor because the wider power spectrum of each subcarrier is less robust to frequency-selectivity. Therefore, there exists an optimum value in N_{SC} to minimize the bit error rate (see Fig. 4.2).

Furthermore, as the guard duration (Δ) increases, the transmission performance becomes poor because the signal transmission in the guard duration in-

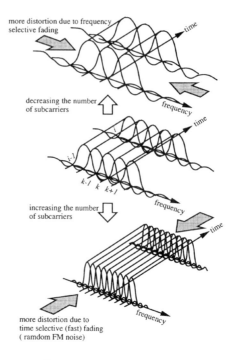

Figure 4.2   Number of subcarriers.

troduces power loss. On the other hand, as Δ decreases, it becomes more sensitive to frequency-selectivity because the shorter guard duration is less robust to delay spread. Therefore, there exists an optimum value in Δ to minimize the bit error rate (BER) (see Fig. 4.3).

For the M–ary DPSK OFDM, we can easily optimize N_{SC} and Δ, because the theoretical BER expression has been obtained in a closed form *for any multipath delay profile and any Doppler power spectrum* [7, 8, 9]. Figs. 4.4 (a) and (b) show the optimum values of the number of subcarriers (N_{SC}) and guard duration (Δ) against the maximum Doppler frequency (f_D) and the RMS delay spread (τ_{RMS}), respectively, where Δ, τ_{RMS} and f_D are normalized by the total transmission rate (R). Here, we assume:

- an exponential-type multipath delay profile with root mean square (RMS) delay spread of τ_{RMS}, which is composed of 20 multipaths,

- a Doppler power spectrum with maximum Doppler frequency of f_D when an omnidirectional monopole antenna is used at the receiver.

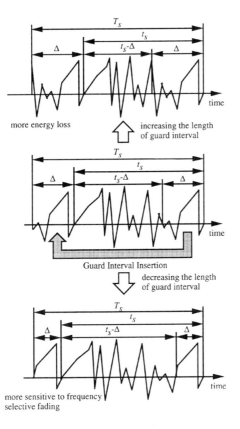

Figure 4.3 Length of guard interval.

3 PILOT-ASSISTED SYNCHRONIZATION APPROACH

3.1 FREQUENCY OFFSET ESTIMATION/COMPENSATION

When a frequency offset is introduced in the radio channel, the BER degrades drastically, since severe inter-subcarrier interference occurs because of the overlapping power spectra between subcarriers (see Fig. 4.6). This sensitivity to frequency offset is often pointed out as a major OFDM disadvantage. Therefore, it is essential to develop a fast and accurate frequency offset estimation/compensation method.

Fig. 4.5 (a) shows the time-frequency format of a signal burst of the two-stage pilot-assisted frequency offset compensation method [9]. The header

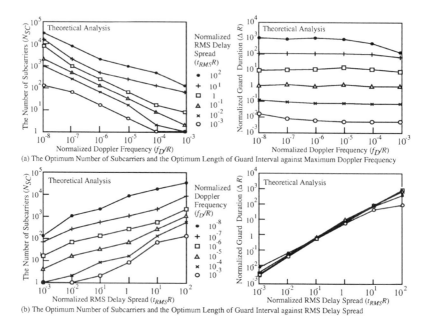

(a) The Optimum Number of Subcarriers and the Optimum Length of Guard Interval against Maximum Doppler Frequency

(b) The Optimum Number of Subcarriers and the Optimum Length of Guard Interval against RMS Delay Spread

Figure 4.4   Optimum number of subcarriers and optimum length of guard interval.

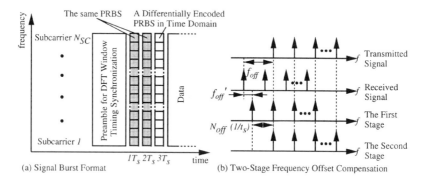

(a) Signal Burst Format   (b) Two-Stage Frequency Offset Compensation

Figure 4.5   Two-stage pilot-assisted frequency offset compensation.

is composed of two parts, a preamble for the DFT window timing synchronization and a group of pilot signals for the carrier frequency synchronization,
where the same pseudo-random binary sequence (PRBS) is inserted at $t = 1 \cdot T_s$
and $t = 2 \cdot T_s$ and a differentially encoded PRBS at $t = 3 \cdot T_s$.

The frequency offset is generally written as

$$f_{off} = N_{off} \cdot (1/t_s) + f'_{off}, \tag{4.5}$$

where N_{off} is an integer and $|f'_{off} \cdot t_s| \leq 0.5$. The first and second stages estimate f'_{off} and N_{off}, respectively.

Assume that the preamble has established perfect DFT window timing synchronization. First, observing the phase shift between the DFT outputs at $t = 1 \cdot T_s$ and $t = 2 \cdot T_s$ of the m–th subcarrier ($\Delta\theta^m$), we can estimate f'^m_{off} as:

$$\begin{aligned} \widehat{f'^m_{off}} &= \frac{1}{2\pi T_s} \Delta\theta^m \\ &= \frac{1}{2\pi T_s} \tan^{-1} \frac{\Im[r_{m,2} r^*_{m,1}]}{\Re[r_{m,2} r^*_{m,1}]}, \end{aligned} \tag{4.6}$$

where $r_{m,i}$ is the DFT output of the m–th subcarrier at $t = i \cdot T_s$ ($i = 1, 2$). Then, averaging $\Delta\theta^m$ ($m = 0, \cdots, N - 1$) over all the subcarriers yields (the first stage)

$$\widehat{f'_{off}} = \frac{1}{2\pi N T_s} \sum_{m=0}^{N-1} \tan^{-1} \frac{\Im[r_{m,2} r^*_{m,1}]}{\Re[r_{m,2} r^*_{m,1}]}. \tag{4.7}$$

Next, using the autocorrelation characteristic of the differentially detected PRBS at $t = 3 \cdot T_s$ (if there are few errors in the PRBS), the second stage can remove the frequency ambiguity N_{off} (see Fig. 4.6 (b)).

Fig. 4.6 shows the BER for the two-stage frequency offset compensation method employing a PRBS with period of 128, where 0 is added to the 7-stage maximum length shift register sequence (the PRBS is used to reduce the envelop peak value). The first stage can compensate for the frequency offset accurately only for $|f'_{off} \cdot t_s| \leq 0.5$, and the second stage can extend the estimation/compensation range $|f_{off} \cdot t_s| \leq 6$ [9].

3.2   SUBCARRIER RECOVERY

For OFDM with M–ary Quadrature Amplitude Modulation (QAM), it is necessary to support coherent detection at the receiver. Instantaneous fading frequency distortion and its time variation can be estimated with periodically-inserted pilot signals. There are two approaches in pilot-assisted subcarrier recovery, namely, frequency domain pilot (FDP) [9] and time domain pilot (TDP) [10].

Fig. 4.7 (a) shows the signal format of the FDP technique. The transmitter inserts a known pilot signal having a maximum amplitude in every N_f subcarriers and N_t symbols. At the receiver, for subcarrier having no pilot signals,

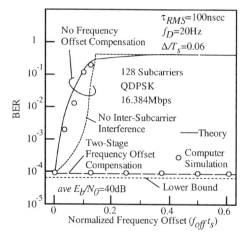

Figure 4.6   Bit error rate performance.

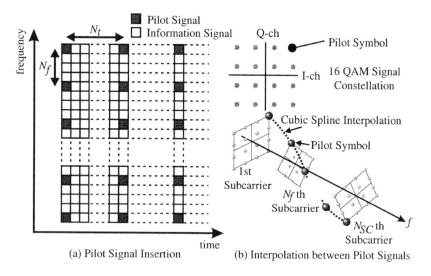

(a) Pilot Signal Insertion   (b) Interpolation between Pilot Signals

Figure 4.7   Frequency domain pilot (FDP) signal insertion.

the reference signals are recovered by *cubic spline*-based interpolation cubic spline-based interpolation (partially, extrapolation) of the received pilot signals (see Fig. 4.7 (b)). The time variation is tracked by linear interpolation of the inserted and estimated pilot signals.

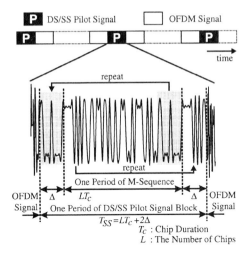

Figure 4.8 Time domain pilot (TDP) signal insertion.

On the other hand, Fig. 4.8 shows the signal format of the TDP technique. The transmitter periodically inserts a Direct Sequence Spread Spectrum (DS/SS) pilot signal in train of OFDM signals. The receiver first estimates instantaneous channel impulse response with the DS/SS pilot signal, and then calculates corresponding instantaneous frequency response essential for subcarrier recovery with the DFT of the estimated instantaneous channel impulse response. The time variation is also tracked by linear interpolation of the estimated amplitude and phase of subcarriers.

Figs. 4.9 (a) and (b) show the BER against the delay spread and the average E_b/N_0, respectively. Here, we assume:

- The number of Subcarriers: 512,

- Transmission Rate: 16.348 Msymbols/sec,

- Guard Interval: $\Delta = 0.1t_s$,

- Filter for DS/SS Signal: Nyquist Filter ($\alpha = 0.5$),

- Multipath Delay Profile: 6-Path Exponentially Decaying.

When the delay spread is small, the FDP outperforms the TDP, however, as the delay spread increases, the performance of the FDP becomes worse, whereas the TDP can keep the performance. In this sense, *the TDP is robust to variation of delay spread.* It is because the TDP can always accurately estimate the channel frequency response as long as the delay spread is within

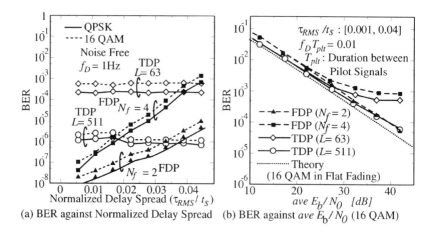

Figure 4.9   Bit error rate performance.

the guard interval, whereas estimation error with interpolation employed in the FDP becomes large as the delay spread increases (variation in frequency response is much larger than that in impulse response).

4   BLIND SYNCHRONIZATION APPROACH

4.1   JOINT FREQUENCY OFFSET/SYMBOL TIMING/SYMBOL PERIOD ESTIMATION

As mentioned in Section 2, in order to maintain orthogonality among sub-carriers even in multipath fading channels, each OFDM symbol is cyclically extended with the guard interval, whose waveform is exactly the same as the tail of the signal itself (see Fig. 4.3). In other words, OFDM system transmits the same waveform twice in each symbol period, so we can deal with it as *"an unknown pilot signal,"* unlike a normal pilot signal whose waveform we know. Therefore, making effective use of the unknown pilot signal, we can estimate frequency offset , symbol timing and symbol period without any extra pilot signal transmission.

The likelihood function on frequency offset, symbol timing and symbol period (DFT window duration) becomes [11] [12]

$$
\lambda(\tau, f_{off}, t_s) = \Re \left[e^{j2\pi f_{off} t_s} \sum_{m=1}^{M} \int_{-\Delta}^{0} r(t + \tau - mT_s) \right.
$$
$$
\left. \times r^*(t + t_s + \tau - mT_s) dt \right]. \tag{4.8}
$$

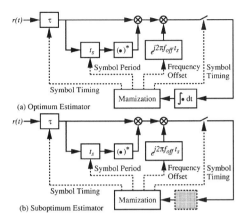

Figure 4.10 Blind frequency offset/symbol timing/symbol period estimation.

The frequency offset f_{off}, the symbol timing τ and the symbol period t_s can be estimated by searching the corresponding values to maximize the likelihood function. Fig. 4.10 (a) shows the block diagram of the estimator. We call the estimator based on Eq. (4.8) the "optimum estimator."

Furthermore, in order to reduce the complexity of the estimator, we can remove the integral operations in Eq. (4.8). The approximated likelihood function is written as

$$\lambda'(\tau, f_\Delta, t_s) = \Re\left[e^{j2\pi f_\Delta t_s} \sum_{m=-\infty}^{\infty} r(t + \tau - mT_s)r^*(t + t_s + \tau - mT_s)\right].$$
(4.9)

Fig. 4.10 (b) shows the block diagram of the estimator, and we call it the "suboptimum estimator."

Figs. 4.11 (a), (b) and (c) show the estimation performance: the RMS frequency error, the RMS symbol timing error and the RMS period error, respectively. Parameters for computer simulation are as follows;

■ The Number of Subcarriers: 32,

■ Modulation/Detection: Differentially-Encoded QPSK/Differential Detection,

■ Symbol Period: 36 samples,

■ Guard Interval: 4 samples,

■ Useful Symbol Period: 32 samples,

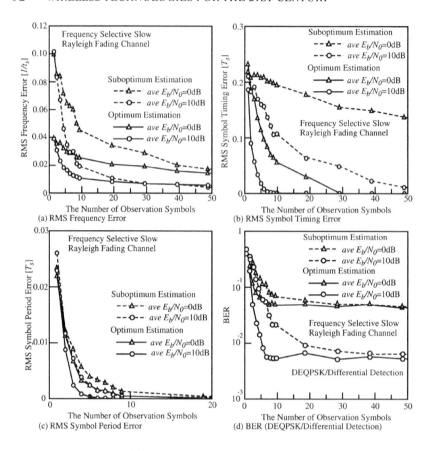

Figure 4.11   Estimation and bit error rate performance.

■ Delay Profile: Exponential (3 samples).

These figures show the excellent of the optimum and suboptimum estimators and the superiority of the optimum estimator over the suboptimum one. The optimum estimator requires only 10-to-20 observation symbols to obtain good estimate for frequency offset, symbol timing and symbol period estimation.

Fig. 4.11 (d) shows the BER performance. This figure clearly shows that 10 observation symbols are sufficient to achieve good BER performance, regardless of values of ave E_b/N_0.

4.2   SUBCARRIER RECOVERY

It is also possible to carry out subcarrier recovery , namely, to estimate the amplitude and phase of each subcarrier, making effective use of the unknown pilot signal.

Assume that the estimator output is optimally sampled with sampling period t_{smp} after frequency offset compensation. The discrete impulse response of the channel is written as

$$c(t) = \sum_{j=0}^{J-1} q_j \delta(t - j t_{smp}),\qquad(4.10)$$

where J is the number of multipaths in the impulse response. The l–th received signal in the i–th symbol period is written as

$$
\begin{aligned}
r_l^i &= r^i(t = l t_{smp}) = (c \otimes s)(t = l t_{smp} - \tau) + z(t = l t_{smp}) \\
&= \sum_{j=0}^{J-1} q_j s^i(l t_{smp} - j t_{smp}) + z(l t_{smp}),\qquad(4.11)
\end{aligned}
$$

where $z(t)$ is the complex-valued white Gaussian noise with zero mean.

Defining M as the number of samples in the guard interval, the received signals in the i–th guard interval ($1 \leq l \leq M$) and the tail of the i–th useful symbol period ($N + 1 \leq l \leq N + M$) are written as (see Fig. 4.12)

$$
M \begin{cases}
r_1^i = q_0 s_1^i + q_1 s_{M+N}^{i-1} + \cdots + q_{J-1} s_{M+N-J+2}^{i-1} + z_1, \\
\quad\vdots \\
r_L^i = q_0 s_L^i + q_1 s_{M-1}^i + \cdots + q_{J-1} s_{M+J+1}^i + z_M,
\end{cases}\qquad(4.12)
$$

$$
M \begin{cases}
r_{N+1}^i = q_0 s_{N+1}^i + q_1 s_N^i + \cdots + q_{J-1} s_{N-J+2}^i + z_{N+1}, \\
\quad\vdots \\
r_{N+M}^i = q_0 s_{N+M}^i + q_1 s_{N+M-1}^i + \cdots + q_{J-1} s_{N+M-J+1}^{i-1} + z_{N+M},
\end{cases}\qquad(4.13)
$$

where $z_l = z(l t_{smp})$. Define the following vectors:

$$
\begin{aligned}
\mathbf{r}_{pre}^i &= [r_1^i, r_2^i, \cdots, r_L^i]^T,\qquad&(4.14) \\
\mathbf{r}_{post}^i &= [r_{N+1}^i, r_{N+2}^i, \cdots, r_{N+L}^i]^T,\qquad&(4.15)
\end{aligned}
$$

where T denotes the transpose. Furthermore, define the following i–th correlation matrix:

$$\mathbf{R}^i = \mathbf{r}_{pre}^i \cdot \mathbf{r}_{post}^{i\ H},\qquad(4.16)$$

where H denotes the Hermitian transpose.

Taking into account the fact:

$$s_1^i = s_{N+1}^i, \quad \cdots \quad, s_M^i = s_{N+M}^i,\qquad(4.17)$$

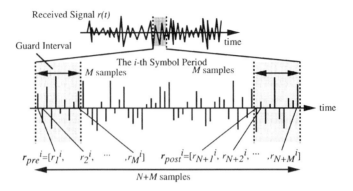

Figure 4.12   Blind subcarrier recovery.

the average correlation matrix can be decomposed as

$$\mathbf{R} =< \mathbf{R}^i >= \frac{1}{I_{av}} \sum_{i=1}^{I_{av}} \mathbf{R}^i = \mathbf{Q}\mathbf{Q}^H \sigma_s^2 + \sigma_n^2 \mathbf{I}, \qquad (4.18)$$

where σ_s^2 and σ_n^2 are the powers of the OFDM signal and the noise, respectively, $\mathbf{I}$ is the $M \times M$ identity matrix, and $\mathbf{Q}$ is the lower triangular matrix written as

$$\mathbf{Q} = \begin{bmatrix} q_0 & & & & & & \\ q_1 & \ddots & & & & & \\ \vdots & \ddots & \ddots & & & 0 & \\ q_{J-1} & & \ddots & \ddots & & & \\ 0 & \ddots & & \ddots & \ddots & & \\ \vdots & \ddots & \ddots & & & \ddots & \ddots \\ 0 & \cdots & 0 & q_{J-1} & \cdots & q_1 & q_0 \end{bmatrix}. \qquad (4.19)$$

Therefore, with the LU factorization of the correlation matrix of the unknown pilot signals, we can estimate the impulse response, and then we can obtain the frequency transfer function essential for subcarrier recovery, namely, the amplitude and phase of each subcarrier from the Fourier Transform of the estimated impulse response.

Fig. 4.13 (a) shows the RMS estimation error of impulse response, where the channel parameters for computer simulation are all the same as in the previous subsection, namely, the case of joint frequency offset/symbol timing/symbol period estimation. Blind estimation of impulse response normally

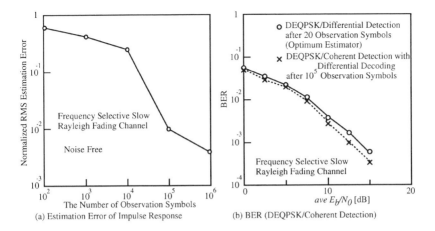

Figure 4.13 Estimation and bit error rate performance.

requires longer observation symbols, and in this case, 10^5 observation symbols are required to obtain good estimate.

Finally, Fig. 4.13 (b) shows the BER performance when DEQPSK /coherent detection with differential decoding is employed. For comparison purpose, the BER for DEQPSK/differential detection is also shown in the figure. Here, the BER is evaluated after 20 observation symbols for differential detection whereas 10^5 symbols for coherent detection. The coherent detection is superior to the differential detection, however, the difference is not so large. However, this figure clearly shows that, without any extra known pilot signal transmission, frequency offset, symbol timing and symbol period and furthermore impulse response for subcarrier recovery can be estimated, making effective use of the unknown pilot signal.

5   CONCLUSIONS

This chapter has presented the principle of the OFDM, and discussed several synchronization techniques essential for successful digital transmission in multipath fading channels.

References

[1] R. W. Chang, "Orthogonal Frequency Division Multiplexing," U. S. Patent3, 488, 445, filed 1966, issued Jan. 6, 1970.

[2] S. B. Weinstein and P. M. Ebert, "Data Transmission By Frequency-Division Multiplexing Using the Discrete Fourier Transform," *IEEE Trans.*

on *Commun. Tech.*, vol. COM-19, no. 5, pp. 628-634, May 1971.

[3] L. J. Cimini, "Analysis and Simulation of a Digital Mobile Channel using Orthogonal Frequency Division Multiplexing," *IEEE Trans. on Commun.*, vol. COM-33, no. 6, pp. 665-675, June 1985.

[4] M. Alard, R. Lassalle, "Principle of Modulation and Channel Coding for Digital Broadcasting for Mobile Receivers," *EBU Technical Review*, no. 224, pp. 168-190, 1987.

[5] B. L. Floch, R. Halbert-Lassalle and D. Castelain, "Digital Sound Broadcasting to Mobile Receivers," *IEEE Trans. on Consumer Electronics*, vol.CE-35, no. 3, pp. 493-503, Mar. 1989.

[6] J. A. C. Bingham, "Multicarrier Modulation for Data Transmission: An Idea Whose Time Has Come," *IEEE Commun. Magazine*, vol.28, no. 5, pp. 5-14, May 1990.

[7] M. Okada, S. Hara and N. Morinaga, "Bit Error Rate Performance of Orthogonal Multicarrier Modulation Radio Transmission Systems," *IEICE Trans. on Comm.*, vol. E76-B, no.2, pp. 113-119, Feb. 1993.

[8] S. Hara, K. Fukui and N. Morinaga, "Multicarrier Modulation Technique for Broadband Indoor Wireless Communications," *Proc. of PIMRC'93*, pp.132-136, Sept. 1993.

[9] S. Hara, M. Mouri, M. Okada and N. Morinaga, "Transmission Performance Analysis of Multi-Carrier Modulation in Frequency Selective Fast Rayleigh Fading Channel," *Wireless Personal Communications*, vol.2, no.4, pp.335-356, 1996.

[10] D. Imamura, S. Hara and N. Morinaga, "Pilot-Assisted Subcarrier Recovery Methods for OFDM Systems (in Japanese)," *IEICE Trans. on Comm.*, vol. J82-B, no. 3, pp. 393-401, Mar. 1999.

[11] M. Okada, M. Mouri, S. Hara, S. Komaki and N. Morinaga, "A Maximum Likelihood Symbol Timing, Symbol Period and Frequency Offset Estimator for Orthogonal Multi-Carrier Modulation Signals," *Proc. of IEEE ICT'96*, pp. 596-601, Apr. 1996.

[12] M. Okada, S. Hara, S. Komaki and N. Morinaga, "Optimum Synchronization of Orthogonal Multi-Carrier Modulated Signals," *Proc. of IEEE PIMRC'96*, pp. 863-867, Oct. 1996.

[13] S. Hara, "Blind Frequency Offset/Symbol Timing/Symbol Period Estimation and Subcarrier Recovery for OFDM Signals in Fading Channels," *Multi-Carrier Spread-Spectrum and Related Topics*, pp. 271-278, Kluwer Academic Publishers, 2000.

Chapter 5

OVERVIEW ON LINEAR MULTIUSER EQUALIZERS FOR DS-CDMA SYSTEMS

Markku Juntti

University of Oulu, Centre for Wireless Communications,
P.O. Box 4500, FIN-90014 University of Oulu, Finland
markku.juntti@ee.oulu.fi

Kari Hooli

University of Oulu, Centre for Wireless Communications,
P.O. Box 4500, FIN-90014 University of Oulu, Finland
kari.hooli@ee.oulu.fi

Abstract   The ideas and basic principles of multiuser receivers for code-division multiple-access (CDMA) systems are summarized and reviewed. Linear multiuser receiver formulation and combination of MAI suppression with multipath and antenna combining is the main contribution of the chapter. The multiuser equalization can be performed either after maximal ration combining or there can be a separate equalizer in each rake receiver branch. If the latter choice is made, the correlation of noise in the equalizer outputs needs to be taken into consideration when selecting the combining weights. Efficient bit error probability (BEP) evaluation methods for such linear receivers with uncoded transmission are also presented. Thirdly, chip equalizers for downlink are described. Their major advantage is that they are applicable also in CDMA systems where the period of spreading sequences is significantly longer than the data symbol period.

Keywords: Multisensor receiver, multiuser receiver, bit error probability, equalizer, code-division multiple-access, receiver design.

1   INTRODUCTION

The application of direct-sequence (DS) spread-spectrum (SS) code-division multiple-access (CDMA) [1, 2, 3, 4] to commercial cellular communications systems has been initialized in cdmaOne (former IS-95) standard. The third generation cellular communication systems, so called International Mobile Telecommunications 2000 (IMT-2000, or, Universal Mobile Telecommunication Systems (UMTS) in Europe, employ various forms of CDMA in their air interface, which is called UMTS Terrestrial Radio Access (UTRA). The radio access techniques include wideband CDMA (W-CDMA), multicarrier based extension of cdmaOne (called cdma2000), and combined time-division code-division multiple-access (TD-CDMA) to be used in time-division duplex (TDD) operation mode. In addition to the cellular applications, CDMA is used as multiaccess technique in local wireless broadband services, like wireless local area networks in ISM band [5, 6].

In CDMA systems several users transmit their signals at the same frequency at the same times. The user transmissions can be identified by their unique signature signals, which are formed by different spreading sequences or spreading codes. The signature signals are usually designed to pose as low crosscorrelation levels as possible. As one extreme the codes can be designed to be totally orthogonal. However, the number of orthogonal spreading sequences is limited to the value of the spreading factor (SF). Therefore, if the number of users or CDMA signals needs to be larger than the value of the SF, all the signature signals cannot be orthogonal. Even if the spreading signals were orthogonal, asynchronous transmissions of different users or different propagation delays in radio channels of various users make the received spreading signals nonorthogonal. For the reasons described above, in most practical cases, there is multiple-access interference (MAI) present in DS-CDMA systems.

Since the conventional correlator receiver is interference limited [4] and suffers a severe performance penalty in DS-CDMA systems, multiuser detection (MUD) has been proposed to improve the receiver performance and CDMA system capacity. Since the optimal multiuser detector [7] has high computational complexity, several suboptimal multiuser detectors have been proposed. See, e.g., [8, 9, 10] for an overview and further references on multiuser detection. A comprehensive textbook treatment is presented in [10]. A brief textbook treatment can be found in [11, Chap. 15].

One important approach to improve the performance of a radio communication receiver is to use multiple antennas to implement spatial diversity or beamforming [12, 13]. It has been a hot topic in recent research, and is currently under practical implementation in several commercial communication systems. The diversity can be either transmit [14] or receive diversity. The former is often applied in downlink (DL) (i.e., forward link) while the latter technique is well suited for uplink (UL) (i.e., reverse link). The idea therein is to have the multiple antenna solution in the base transceiver station (BTS) in both cases. The beamforming is also implemented by BTS antenna arrays both for UL and DL. In CDMA systems, the use of multiple antennas is particularly attractive, since it reduces the required transmit power levels, which directly increases the system capacity [4].

The purpose of this chapter is threefold. First, the ideas and basic principles of multiuser receivers are briefly summarized and reviewed; the various techniques are formulated in a unique framework. These issues, and the definition of the system model, are the topics of Section 2. Secondly, the main part of the paper, presented in Section 3, consists of describing the linear multiuser receiver formulation and combination of MAI suppression with multipath and antenna combining. Efficient bit error probability (BEP) evaluation methods for such linear receivers with uncoded transmission are also described. Thirdly, a recent development on the potential application of multiuser receiver principles in downlink is described in Section 5. In Section 6, the chapter is summarized and conlusions are drawn.

2   PRELIMINARIES

2.1   SYSTEM MODEL

The system model is defined in this section to set up the further notations used later in the treatment. The conventional CDMA signal description given herein is at its best when uplink (a multiple-access channel) of a typical CDMA system is considered.

The CDMA system is assumed to include K active users whose transmissions are received by M different antennas[1], or, more generally, sensors. The received CDMA signal in mth antenna, where $m \in \{1, 2, \ldots, M\}$, is the convolution of the transmitted signal and the channel impulse response plus the additive channel noise. Thus, the complex envelope of the received signal in antenna m can be expressed as

$$r_m(t) = \sum_n \sum_{k=1}^{K} b_k^{(n)} A_{k,m} \sum_{l=1}^{L} c_{k,l,m}^{(n)} s_k(t - nT - \tau_{k,m} - \tau_{k,l,m}) + \eta_m(t), \quad (5.1)$$

where n is the discrete time index referring to the symbol interval, K is the number of users in the CDMA system, L is the number of propagation paths in the channel (assumed equal for all users and antennas for notational sim-

---

[1]The antennas may be closely spaced antenna elements used in adaptive antenna arrays to perform beamforming, or they may be widely spaced antenna elements used to provide receive diversity. The main emphasis in this paper is given to the diversity case, but the considered receiver principles themselves can be equally applied to antenna arrays as well.

plicity), $b_k^{(n)} \in \Xi$ is the transmitted data symbol at symbol interval n (i.e., as $t \in [nT, (n+1)T))$, Ξ is the modulation symbol alphabet, $A_{k,m} = \sqrt{E_{k,m}}$ is the received amplitude of user k in antenna m, $E_{k,m}$ is the energy per symbol of the corresponding real bandpass signal in antenna m, $\tau_{k,m} \in [0, T)$ is the delay of kth user's received signal in antenna m, $c_{k,l,m}^{(n)}$ is the complex gain (includes both the impact of Rayleigh fading and direction of arrival) and $\tau_{k,l,m} \in [0, T_m]$ is the delay of the lth multipath component of user k on symbol interval n in antenna m, T_m is the delay spread of the multipath channel, $\eta_m(t)$ is complex zero mean additive white Gaussian noise (AWGN) process in antenna m with two-sided power spectral density σ^2 (assumed equal for all antennas for notational simplicity without loss of generality), and $s_k(t)$ is the signature signal[2] of user k. Since AWGN in the received signal is mostly due to the radio frequency front-end of the antennas, the noise processes $\eta_m(t)$ in each antenna can usually be assumed to be independent of each other. For convenience $s_k(t)$ is assumed to be normalized so that $s_k(t) = 0$, if $t \notin [0, T)$, and $\int_0^T |s_k(t)|^2 \, dt = 1$.

It is easy to see that the matched filter (MF) outputs of all antennas for all users and multipath components produce sufficient statistics for the detection of the data symbols. The sampled output of the matched filter of the kth users lth multipath component in antenna m on symbol interval n is

$$y_{k,l,m}^{(n)} = \int_{nT+\tau_{k,m}+\tau_{k,l,m}}^{(n+1)T+\tau_{k,m}+\tau_{k,l,m}} r_m(t) s_k(t - nT - \tau_{k,m} - \tau_{k,l,m}) dt.$$

Let the vectors of MF output samples in antenna m for symbol interval n be defined as

$$\mathbf{y}_{k,m}^{(n)} = (y_{k,1,m}^{(n)}, y_{k,2,m}^{(n)}, \ldots, y_{k,L,m}^{(n)})^{\mathrm{T}} \in \mathbb{C}^L,$$

$$\mathbf{y}_m^{(n)} = ((\mathbf{y}_{1,m}^{(n)})^{\mathrm{T}}, (\mathbf{y}_{2,m}^{(n)})^{\mathrm{T}}, \ldots, (\mathbf{y}_{K,m}^{(n)})^{\mathrm{T}})^{\mathrm{T}} \in \mathbb{C}^{KL}.$$

In the sequel the signal is considered over an observation window of finite length $N = 2P + 1$ [17] so that the symbol interval of interest is set to $n = 0$. The concatenation of MF outputs over the observation window is

$$\boldsymbol{y}_m = \left((\mathbf{y}_m^{(-P)})^{\mathrm{T}}, \ldots, (\mathbf{y}_m^{(0)})^{\mathrm{T}}, \ldots, (\mathbf{y}_m^{(P)})^{\mathrm{T}} \right)^{\mathrm{T}} \in \mathbb{C}^{NKL}.$$

The vector of the matched filter outputs has expression

$$\boldsymbol{y}_m = \mathcal{R}_m \mathcal{C}_m \mathcal{A}_m \boldsymbol{b} + \boldsymbol{n}_m = \mathcal{R}_m \boldsymbol{h}_m + \boldsymbol{n}_m, \tag{5.2}$$

---

[2] It has been assumed that the signature signals are periodic over the symbol interval. This assumption is due to notational convenience only. It is easy to generalize to the case of longer signature signals [15, 9]. Similarly, the assumption of a single spreading factor or processing gain for all users can be straightforwardly generalized [16].

where

$$
\mathcal{R}_m = \begin{pmatrix}
\mathbf{R}_m(0) & \mathbf{R}_m^\mathrm{T}(1) & \mathbf{R}_m^\mathrm{T}(2) & \cdots & \mathbf{0}_K \\
\mathbf{R}_m(1) & \mathbf{R}_m(0) & \mathbf{R}_m^\mathrm{T}(1) & \cdots & \mathbf{0}_K \\
\mathbf{R}_m(2) & \mathbf{R}_m(1) & \mathbf{R}_m^\mathrm{T}(0) & \cdots & \mathbf{0}_K \\
\vdots & \vdots & \vdots & & \vdots \\
\mathbf{0}_K & \mathbf{0}_K & \mathbf{0}_K & \cdots & \mathbf{R}_m(0)
\end{pmatrix} \in \mathbb{R}^{NKL \times NKL},
$$

$$
\mathbf{R}_m(i) = \begin{pmatrix}
\mathbf{R}_{1,1,m}(i) & \mathbf{R}_{1,2,m}(i) & \cdots & \mathbf{R}_{1,K,m}(i) \\
\mathbf{R}_{2,1,m}(i) & \mathbf{R}_{2,2,m}(i) & \cdots & \mathbf{R}_{2,K,m}(i) \\
\vdots & \vdots & & \vdots \\
\mathbf{R}_{K,1,m}(i) & \mathbf{R}_{K,2,m}(i) & \cdots & \mathbf{R}_{K,K,m}(i)
\end{pmatrix} \in \mathbb{R}^{KL \times KL},
$$

matrices $\mathbf{R}_{k,k',m}(i) \in \mathbb{R}^{L \times L}$ have elements

$$
\left(R_{k,k',m}(i) \right)_{l,l'} = \int_{-\infty}^{\infty} s_k(t - \tau_{k,m} - \tau_{k,l,m}) s_{k'}(t + iT - \tau_{k',m} - \tau_{k',l',m}) dt,
$$

$$
\mathcal{C}_m = \mathrm{diag}\left(\mathbf{C}_m^{(n-P)}, \mathbf{C}_m^{(n-P+1)}, \ldots, \mathbf{C}_m^{(n+P)} \right)^\mathrm{T} \in \mathbb{C}^{NKL \times NK},
$$

$$
\mathbf{C}_m^{(n)} = \mathrm{diag}\left(\mathbf{c}_{1,m}^{(n)}, \mathbf{c}_{2,m}^{(n)}, \ldots, \mathbf{c}_{K,m}^{(n)} \right) \in \mathbb{C}^{KL \times K},
$$

$$
\mathbf{c}_{k,m}^{(n)} = \left(c_{k,1,m}^{(n)}, c_{k,2,m}^{(n)}, \ldots, c_{k,L,m}^{(n)} \right)^\mathrm{T} \in \mathbb{C}^L,
$$

$$
\mathcal{A}_m = \mathrm{diag}\left(\mathbf{A}_m, \mathbf{A}_m, \ldots, \mathbf{A}_m \right) \in \mathbb{C}^{NK \times NK},
$$

$$
\mathbf{A}_m = \mathrm{diag}\left(A_{1,m}, A_{2,m}, \ldots, A_{K,m} \right) \in \mathbb{C}^{K \times K},
$$

$$
\boldsymbol{b} = \left((\mathbf{b}^{(-P)})^\mathrm{T}, \ldots, (\mathbf{b}^{(0)})^\mathrm{T} \ldots, (\mathbf{b}^{(P)})^\mathrm{T} \right)^\mathrm{T} \in \Xi^{NK},
$$

$$
\mathbf{b}^{(n)} = \left(b_1^{(n)}, b_2^{(n)}, \ldots, b_K^{(n)} \right)^\mathrm{T} \in \Xi^K,
$$

$\boldsymbol{h}_m = \mathcal{C}_m \mathcal{A}_m \boldsymbol{b}$ is the data-amplitude product vector, and $\boldsymbol{n}_m$ is the noise vector at the MF outputs. Equation (5.2) can be expressed also in the form

$$
\boldsymbol{y}_m = \mathcal{R}_m \mathcal{B}_m \boldsymbol{c}_m + \boldsymbol{\eta}_m, \tag{5.3}
$$

where

$$
\boldsymbol{c}_m = ((\mathbf{c}_m^{(-P)})^\mathrm{T}, \ldots (\mathbf{c}_m^{(0)})^\mathrm{T}, \ldots, (\mathbf{c}_m^{(P)})^\mathrm{T})^\mathrm{T}, \tag{5.4}
$$

is the vector of complex channel tap gains in the antenna m,

$$
\mathbf{c}_m^{(n)} = ((\mathbf{c}_{1,m}^{(n)})^\mathrm{T}, (\mathbf{c}_{2,m}^{(n)})^\mathrm{T}, \ldots, (\mathbf{c}_{K,m}^{(n)})^\mathrm{T})^\mathrm{T},
$$

$$
\mathcal{B}_m = \mathrm{diag}\left(A_{1,m} b_1^{(-P)} \mathbf{I}_L, A_{2,m} b_2^{(-P)} \mathbf{I}_L, \ldots, A_{K,m} b_K^{(-P)} \mathbf{I}_L, \right.
$$

$$A_{1,m} b_1^{(-P+1)} \mathbf{I}_L, \ldots, A_{K,m} b_K^{(P)} \mathbf{I}_L \Big) \in \mathbb{C}^{NKL \times NKL} \qquad (5.5)$$

is a matrix of products of the data symbols and average amplitudes, and $\mathbf{I}_L$ is an identity matrix of size $L \times L$.

To combine the observations to a single vector equation, let

$$\boldsymbol{y} = (\boldsymbol{y}_1^T, \boldsymbol{y}_2^T, \ldots, \boldsymbol{y}_M^T) \in \mathbb{C}^{MNKL} \qquad (5.6)$$

is the vector of concatenated MF outputs of all the antennas. By Eq. (5.3) it can be expressed in the form

$$\boldsymbol{y} = \mathcal{R}\mathcal{C}\mathcal{A}\boldsymbol{b} = \mathcal{R}\mathcal{B}\boldsymbol{c} + \boldsymbol{\eta}, \qquad (5.7)$$

where $\mathcal{R} = \mathrm{diag}(\mathcal{R}_1, \mathcal{R}_2, \ldots, \mathcal{R}_M) \in \mathbb{C}^{MNKL \times MNKL}$, $\mathcal{C} = \mathrm{diag}(\mathcal{C}_1, \mathcal{C}_2, \ldots, \mathcal{C}_M) \in \mathbb{C}^{MNKL \times MNK}$, $\mathcal{A} = (\mathcal{A}_1, \mathcal{A}_2, \ldots, \mathcal{A}_M)^T \in \mathbb{R}^{MNK \times NK}$, $\mathcal{B} = \mathrm{diag}(\mathcal{B}_1, \mathcal{B}_2, \ldots, \mathcal{B}_M) \in \mathbb{C}^{MNKL \times MNKL}$, $\boldsymbol{c} = (\boldsymbol{c}_1^T, \boldsymbol{c}_2^T, \ldots, \boldsymbol{c}_M^T) \in \mathbb{C}^{MNKL}$, and $\boldsymbol{\eta} = (\boldsymbol{\eta}_1^T, \boldsymbol{\eta}_2^T, \ldots, \boldsymbol{\eta}_M^T) \in \mathbb{C}^{MNKL}$.

2.2   OPTIMAL MULTIANTENNA MULTIUSER RECEIVERS

It is relatively straightforward to show that the optimal multiuser maximum-likelihood sequence detector (MLSD) (without forward error control coding) [7] with known channel parameters, obeys the decision rule [18]

$$\hat{\boldsymbol{b}} = \arg \max_{\boldsymbol{\xi} \in \Xi^{NK}} \Omega(\boldsymbol{\xi}), \qquad (5.8)$$

where $\boldsymbol{\xi}$ is the optimization variable vector,

$$\Omega(\boldsymbol{\xi}) = 2\mathrm{Re}(\boldsymbol{\xi}^H \boldsymbol{u}) - \boldsymbol{\xi}^H \mathcal{H} \boldsymbol{\xi} \qquad (5.9)$$

is the log-likelihood function,

$$\boldsymbol{u} = \sum_{m=0}^{M} \mathcal{A}_m^H \mathcal{C}_m^H \boldsymbol{y}_m = \mathcal{A}^H \mathcal{C}^H \boldsymbol{y} = \mathcal{H}\boldsymbol{b} + \tilde{\boldsymbol{\eta}} \qquad (5.10)$$

is a vector of maximal-ratio combined (MRC) antenna outputs,

$$\mathcal{H} = \sum_{m=0}^{M} \mathcal{A}_m^H \mathcal{C}_m^H \mathcal{R}_m \mathcal{C}_m \mathcal{A}_m, \qquad (5.11)$$

is ,the correlation matrix between the signal components in $\boldsymbol{u}$, and $\tilde{\boldsymbol{\eta}} = \sum_{m=0}^{M} \mathcal{A}_m^H \mathcal{C}_m^H \boldsymbol{\eta}_m$ is the combined noise vector with covariance matrix $\boldsymbol{\Sigma}_{\tilde{\boldsymbol{\eta}}} = \sigma^2 \mathcal{H}$. The optimal MLSD is prohibitively complex for most applications, since the number of required operations depends exponentially on the number of users.

2.3   LINEAR MULTIUSER EQUALIZERS

Due to the complexity of the optimal MLSD receiver, several suboptimal receivers have been proposed and studied. Most of the suboptimal receivers make their decisions as follows

$$\hat{b} = \text{sgn}\left(u_{[MUD]}\right), \tag{5.12}$$

where $u_{[MUD]}$ is the multiuser detector or MAI suppressor output. The output $u_{[MUD]}$ is obtained as an approximation to the optimization problem in Eq. (5.8). The simplest approximation totally neglects the presence of multiple users, i.e., it assumed that $\mathcal{H}$ is a diagonal matrix. The assumption results in the conventional 2-dimensional rake receiver, i.e., $u_{[MUD]} = u$. The conventional receiver can be seen as a special case of linear multiantenna multiuser equalizers, which are discussed in more detail in Section 3.

The other linear multiantenna multiuser equalizers can be obtained with similar reasonings. A simple approximation is to perform the optimization Eq. (5.8) over the set of complex numbers instead of the symbol alphabet, i.e.,

$$u_{[MUD]} = \arg \max_{\xi \in \mathbb{C}^{NK}} \Omega(\xi), \tag{5.13}$$

where ξ is the optimization variable vector. Since $\Omega(\xi)$ is a quadratic function of ξ, it is straightforward to show that the optimum solution is

$$\mathcal{H} u_{[MUD]} = u \Leftrightarrow u_{[MUD]} = \mathcal{H}^{-1} u. \tag{5.14}$$

The result in Eq. (5.14) is known as a zero-forcing (ZF) or decorrelating solution [19, 20]. Another well-known solution, the linear minimum mean squared error (LMMSE) equalizer [21], is obtained, if matrix $\mathcal{H}$ in Eq. (5.9) is replaced by $\mathcal{H} + \sigma^2 \mathbf{I}_{NK}$ in optimization Eq. (5.13). The choice may seem somewhat abrupt, but it will become more clear in Section 3.

2.4   INTERFERENCE CANCELLATION RECEIVERS

In addition to the linear equalizers, another important class of multiuser receivers are the interference cancellation (IC) receivers. When deriving an IC receiver, the main underlying assumption is that the signals of the interfering users are known, and the optimization in Eq. (5.13) is performed for single user only[3] [24]. This leads to the following optimization

$$(u_{[IC]})_i = \arg \max_{\substack{(\xi)_i \in \Theta \\ (\xi)_j = (u)_j \\ j \neq i}} \Omega(\xi), \tag{5.15}$$

where Θ is the set over which the optimization is performed. It follows from Eq. (5.15) that the IC receivers subtract the interference term from the received signal. Since the interference term is in practice unknown, it must be estimated. Thus, in IC receivers MAI term is explicitly estimated and then it is subtracted from the received signal or from the combined signal u. If the interference estimation is successful, the receiver performance is improved due to reduced interference level. A typical feature of the IC receivers is the fact that the interference cleaned signal may be re-used to form new, hopefully improved, interference estimate, which can be used to cancel MAI again (or perform optimization in Eq. (5.15) again). This principle results in an iterative IC receiver, which has often been called multistage interference cancellation receiver.

Conceptually the simplest IC receiver is probably the multistage parallel interference cancellation (PIC) receiver [25, 26, 27, 24, 28, 29, 30, 31, 32]. The output at the mth stage can be presented as [24]

$$u_{[PIC]}(m) = u - \hat{\Psi}_{[PIC]}(m), \tag{5.16}$$

where $\Psi(m)$ is the interference estimate. It has been formed based on the output of stage $m - 1$, i.e., $\Psi_{[PIC]}(m) = f(u_{[PIC]}(m-1))$, where f denotes for the interference estimation function (there are more details below). In the first stage the conventional two-dimensional rake receiver or any linear equalizer may be used. In the former case $\Psi_{[PIC]}(m) = f(u)$.

Equation (5.16) describes a Jacobi-type iteration [33]. Another popular IC receiver, namely serial interference cancellation (SIC) receiver [34, 35] is obtained, if the Jacobi iteration is replaced by a Gauss-Seidel iteration. It means that the newest MAI estimates for the symbols estimated in stage m are based on decisions made in stage m. More precisely, the SIC receiver output is

$$(u_{[SIC]})_i(m) = (u)_i - \hat{\Psi}_{[SIC]i}(m), i = 1, 2, \ldots, NK \tag{5.17}$$

where $\hat{\Psi}_{[SIC]i}(m) = f\big((u_{[SIC]}(m))_{1:i-1}, (u_{[SIC]}(m-1))_{i+1:NK}\big)$ is the interference estimate for the ith term in u, and $u_{i:j}$ denotes the elements

---

[3]The idea is very similar to that of the expectation-maximization (EM) or the space alternating generalized EM (SAGE) algorithms [22, 23].

$i, i + 1, \ldots, j$ of vector u. In SIC receivers, the users are usually organized according to their received power levels so that the users with the largest received power are detected and canceled first.

The PIC and SIC receiver can also be combined into a hybrid receiver, where users are grouped according to their power levels [16]. In the resulting group-wise serial interference cancellation (GSIC) parallel cancellation can be used within groups [36, 37, 38]. The grouping can also be used have a trade-off between implementation complexity of the optimal MLSD receiver and performance of the suboptimal receiver. In other words, Eq. (5.15) can be modified so that only part of the interfering signals is assumed known and thus canceled, and the optimization is performed over signals of several users [39, 40]. Since vector u includes symbols from several symbol intervals, similar trade-off can be made also in time-domain [41]. In other words, optimization can Eq. (5.15) can be performed over several consequtive symbols of one user.

The cancellation order distinguishes the parallel and serial cancellation from each other. Another major feature which divides the IC receivers to various classes is the technique used in interference estimation. The main alternatives are so called soft decision (SD) [42, 43] and hard decision (HD) based MAI estimation tecniques. In the latter one, explicit data decisions are made, and MAI estimate is obtained as a product of channel estimates and data decisions. It means that the set Θ, over which the optimization Eq. (5.15) is performed, equals symbol alphabet, i.e., $\Theta = \Xi$. For example, in the case of a HD-PIC receiver, the MAI estimate becomes

$$\Psi_{[HD-PIC]}(m) = \hat{\mathcal{H}}(m-1)\hat{b}(m-1), \qquad (5.18)$$

where $\hat{\mathcal{H}}(m-1)$ and $\hat{b}(m-1)$ denote the tentative channel and data estimates provided by the stage $m-1$ of the multistage HD-PIC receiver. In the soft-decision based parallel interference cancellation the amplitude-data product is estimated linearly without making an explicit data decision. Thus, $\Theta = \mathbb{C}$. In other words, for the SD-PIC receiver, the product $\mathcal{H}(m-1)b(m-1)$ is estimated jointly and the MAI estimate becomes

$$\Psi_{[SD-PIC]}(m) = \widehat{\mathcal{H}b}(m-1) = u_{[SD-PIC]}(m-1). \qquad (5.19)$$

Based on the discussion above, it is easy to see that all the IC receivers are actually iteratively solving the matrix equation

$$\mathcal{H}u_{[IC]} = u. \qquad (5.20)$$

The SD-PIC is using Jacobi iterations and the SD-SIC Gauss-Seidel iterations. The same applies to their HD counterparts if it is noted that they add nonlinear operations to the iterations. By comparing Eq. (5.20) to Eq. (5.14), it is easy to understand that the SD based IC algorithms, if they converge, converge actually to the ZF solution [44]. Similarly, if matrix $\mathcal{H}$ in SD-IC algorithms is replaced by $\mathcal{H} + \sigma^2 \mathbf{I}_{NK}$ (as in Section 2.3), they converge to the LMMSE solution [45]. Other iterative implementations, like the steepest descent and conjugate gradient algorithms, of linear equalizers in [46] are rather similar to the SD-PIC receivers with different iterative algorithms. For HD-IC receivers the relationship to the linear equalizers is not as clear and simple; it is an interesting topic for future research.

3   LINEAR MULTIANTENNA MULTIUSER EQUALIZATION

In this section, the linear multiantenna multiuser equalizers are described from a more conventional perspective as in Section 2.3. The goal is to get more insight into the detailed algorithms and bit error probability analysis. There are several possible architectures for suboptimal multiantenna multiuser receivers applying linear interference suppression filtering [10, 47, 48]. The basic options are to have a separate multiuser equalizer for each antenna element or a single equalizer for the maximal ratio combined signal. The latter is considered in Section 3.1, and the former in Section 3.2.

3.1   COMBINING BEFORE EQUALIZATION

If the equalization is performed based on the maximal ratio combined signal u, linear equalization remains to be a conceptually simple extension of the well-known linear multiuser equalization techniques to the model in Eqs. (5.8)–(5.11). The technique is conceptually simple, since the multipath and antenna combining are not affected by the introduction of the equalization. The equalization principles are most straightforward to describe if so called block equalization approach is used. It means that a whole block of received signal is processed once in the receiver. The block equalization is described in Section 3.1.1. Extension of the principles to filtering based processing is presented in Section 3.1.2.

3.1.1   Block Equalization.   The general idea of linear multiuser equalizers is to process the equalizer input by some linear operation. The output a linear block equalizer can be described as

$$u_{LIN} = T^{\mathrm{H}}u = T^{\mathrm{H}}\mathcal{H}b + \tilde{\eta}_{LIN}, \qquad (5.21)$$

where $T \in \mathbb{C}^{NK \times NK}$ is the matrix describing the block equalizer, and $\tilde{\eta}_{LIN} = T^{\mathrm{H}}\tilde{\eta}$ is the noise term at the equalizer output. The matrix $T^{\mathrm{H}}\mathcal{H}$ describes the convolution of the overall multiuser channel (including multipath propagation as well as antenna and multipath combining) and the equalizer.

Different equalizer designs are obtained by different choices of the equalizer matrix $\mathcal{T}$. The well-known choice for $\mathcal{T}$ include the zero-forcing equalizer [20, 47, 49, 50, 51] $\mathcal{T}_{ZF} = \mathcal{H}^{-1}$, which completely suppresses MAI. The output of the ZF equalizer is by Eq. (5.21) $u_{LIN} = b + \tilde{\eta}_{ZF}$. Although MAI is removed completely by the ZF equalizer, it has the drawback that the noise is enhanced, i.e., the variance of the AWGN term is increased. The noise enhancement problem can be alleviated by the introduction of linear minimum mean squared error equalizer [21, 52, 53] $\mathcal{T}_{MMSE} = (\mathcal{H} + \sigma^2 \mathbf{I}_{NK})^{-1}$. The LMMSE equalizer is a compromise between the conventional receiver and the zero-forcing equalizer. In addition to the ZF and LMMSE equalizers, adaptive linear receivers to minimize the bit error probability [54, 55] have also been proposed.

3.1.2   Filtering Based Equalization.   The block equalization described in the above section is useful for illustrating the basic principles of the equalizer design. It is, however, sometime more practical to implement the equalizers as linear filters. That is possible by taking one column of the block equalizer matrix $\mathcal{T}$ [20], and using the elements of the column as FIR filter tap gains. In an ideal implementation, the size of matrix $\mathcal{T}$, and, thus the length of the corresponding FIR filter equals the number of users times the data packet length. In other words, the window size N would be infinite in an asynchronous CDMA system. The problem can be overcome simply by truncating the equalizer filters, and using only the middle columns of matrix $\mathcal{T}$ [15, 17]. Although such an approach may sound abrupt, it can be shown that the loss in performance becomes marginal with very moderate equalizer lengths [15, 17]. The FIR implementation is convenient in the sense that it gives the opportunity to implement the linear equalizers as simple adaptive filters, if the spreading signals of the users are periodic with short enough periods [56, 57, 58, 59, 60].

As noted in Section 2.4, the SD based IC receivers converge to the ZF solution. It is well known that IC receivers can be implemented with finite window size, which increases with increasing number of cancellation stages [24]. Thus, the FIR type linear equalizers could be implemented by using a SD-IC approach.

3.2   EQUALIZATION BEFORE COMBINING

The equalization after the multipath and antenna combining is conceptually simple, and for that reason an attractive technique. Furthermore, it is the Right Thing To Do in the sense that MRC output vector u is the minimal sufficient statistics for the detection of the data symbols. Unfortunately, the approach has also two major drawbacks. Both of them are caused by the fact that the equalizer impulse response matrix $\mathcal{T}$ depends on the instantaneous channel realizations of all users. The first drawback is the fact that once a channel impulse response of only one user is changed, the whole equalizer matrix needs to be updated. The second drawback is the fact that the equalization provides no gain in channel estimation, since the channels need to be estimated from signals with MAI. The first drawback may make even the simple adaptive implementations useless if the fading rate is even moderate [48, 61, 62, 63]. Therefore, it may sometimes be preferable to have a separate equalizer for each rake branch of each antenna in stead of a single equalizer for the maximal ratio combined signal as in Section 3.1. In this section, the approach of having separate equalizer for each rake branch of each antenna is considered.

A linear multiuser equalizer for kth user's lth multipath component in antenna m is characterized by a FIR filter. The coefficients of the equalizer are included in vector $\boldsymbol{w}_{k,l,m} \in \mathbb{C}^{NKL}$. The filter output is

$$v_{k,l,m} = \boldsymbol{w}_{k,l,m}^{\mathrm{H}} \boldsymbol{y}_m. \tag{5.22}$$

The decision rule for any linear multiuser receiver can be expressed as

$$\hat{b}_k^{(0)} = \mathrm{sgn}(\mathbf{g}_k^{\mathrm{H}} \mathbf{v}_k), \tag{5.23}$$

where

$$\mathbf{v}_k = (v_{k,1,1}, v_{k,2,1}, \ldots, v_{k,L,1}, v_{k,1,2}, \ldots, v_{k,L,M}) \in \mathbb{C}^{ML} \tag{5.24}$$

is a vector containing the kth user's linear interference suppression filter outputs for all antennas and multipath components, and $\mathbf{g}_k \in \mathbb{C}^{ML}$ is the vector with similar structure as $\mathbf{v}_k$ containing the combining coefficients for the multipaths and antennas of user k.

The conventional two-dimensional rake receiver chooses the impulse response $\boldsymbol{w}_{k,l,m}$ as $(kKL + l)$th column of matrix $\mathbf{I}_{NKL}$. The impulse response $\boldsymbol{w}_{k,l,m}$ of a ZE equalizer is the $(kKL + l)$th column of matrix $\mathcal{R}_m^{-1}$. The LMMSE equalizer is the $(kKL + l)$th column of matrix $(\mathcal{R}_m + \sigma^2 \Sigma_{\boldsymbol{h}_m})^{-1}$, where $\Sigma_{\boldsymbol{h}_m}$ is the covariance matrix of the data-amplitude product vector $\boldsymbol{h}_m$ [48, 64].

The equalizer output vector for user k can be expressed in the form

$$\mathbf{v}_k = \mathcal{W}_k^{\mathrm{H}} \boldsymbol{y}, \tag{5.25}$$

where $\mathcal{W}_k = \mathrm{diag}(\mathcal{W}_{k,1}, \mathcal{W}_{k,2}, \ldots, \mathcal{W}_{k,M}) \in \mathbb{C}^{MNKL \times ML}$ is the matrix of equalizer impulse responses for all antenna composed from equalizer impulse response matrices for different antennas of the form $\mathcal{W}_{k,m} = (\boldsymbol{w}_{k,1,m}, \boldsymbol{w}_{k,2,m}, \ldots, \boldsymbol{w}_{k,L,m}) \in \mathbb{C}^{NKL \times L}$.

The kth user's equalizer output vector $\mathbf{v}_k$ can be decomposed into the form

$$\mathbf{v}_k = \mathbf{v}_{[S+MAI]k}(\boldsymbol{b})\boldsymbol{c} + \mathbf{v}_{[N]k}, \tag{5.26}$$

where $\mathbf{v}_{[S+MAI]k}(\mathbf{b})$ is the component including both the desired signal and the remaining MAI components of $\mathbf{v}_k$ excluding the impact of Rayleigh fading, and $\mathbf{v}_{[N]k}$ is the AWGN response at the equalizer output; the latter term has a zero mean and covariance matrix $\Sigma_{\mathbf{v}_{[N]k}}$. The signal+MAI component can be expressed in the form

$$\mathbf{v}_{[S+MAI]k}(\mathbf{b}) = \mathcal{W}_k^{\mathrm{H}} \mathcal{R} \mathcal{B}, \tag{5.27}$$

and the noise covariance is

$$\Sigma_{\mathbf{v}_{[N]k}} = \sigma^2 \mathcal{W}_k^{\mathrm{H}} \mathcal{R} \mathcal{W}_k \in \mathbb{R}^{ML \times ML}. \tag{5.28}$$

The desired signal component of $\mathbf{v}_k$ is obtained from $\mathbf{v}_{[S+MAI]k}(\mathbf{b})$ by substituting $\mathbf{b} = \mathbf{0}$ excluding the element $b_k^{(n)}$. The MAI component $\mathbf{v}_{[MAI]k}(\mathbf{b})$ is obtained from $\mathbf{v}_{[S+MAI]k}(\mathbf{b})$ by substituting $b_k^{(n)} = 0$.

The optimal choice for the combining weight vector is

$$\mathbf{g}_k = \Sigma_{\mathbf{v}_{[MAI+N]k}}^{-1} \Gamma_k \mathbf{c}_k, \tag{5.29}$$

where $\Sigma_{\mathbf{v}_{[MAI+N]k}} = \Sigma_{\mathbf{v}_{[MAI]k}} + \Sigma_{\mathbf{v}_{[N]k}}$ is the covariance of MAI plus noise. Matrix $\Gamma_k \in \mathbb{C}^{ML \times ML}$ is a diagonal matrix including the magnitudes of the multipath components of the desired user k at the outputs of equalizers, i.e.,

$$\Gamma_k = \mathrm{diag}(\mathbf{v}_{[S+MAI]}(\tilde{\mathbf{b}})) \in \mathbb{C}^{ML \times ML}, \tag{5.30}$$

where $\tilde{\mathbf{b}}$ is a vector of all zeros except the element $b_k^{(n)} = 1$. The multiplication by the inverse noise covariance $\Sigma_{\mathbf{v}_{[MAI+N]k}}^{-1}$ in Eq. (5.29) is required to match the MRC to the noise statistics [47]. The multiplication by matrix Γ_k in Eq. (5.29) gives the outputs of each antenna the correct weight which depends on the magnitude of the equalizer output. In other words, matrix Γ_k takes into consideration the possible bias in the equalizer output of a user and the average power differences between antennas $A_{k,m}$; the latter are described by amplitudes $A_{k,m}, m = 1, 2, \ldots, M$ included in matrix Γ_k. The main lesson learned herein is as follows: if a separate equalizer is applied for each multipath component and antenna element, the gains of maximal ratio combining performed on the equalizer outputs are affected by the equalization, except some special cases.

4   BIT ERROR PROBABILITY ANALYSIS

The bit error probability of linear multiuser receivers is analyzed in this section. It is a conceptually simple task in AWGN channels. The analysis is based on that presented originally in [65, 66, 67, 68]. The results are derived for BPSK modulation. Since the BEP for QPSK modulation is approximately the same as for the BEP for BPSK modulation [11, Eq. (5-2-59), p. 272] (assuming that E_k denotes the energy per symbol not energy per bit), the analysis below can be applied to get the approximate BEP for QPSK modulated signals as well.

The kth user's average bit error probability of a linear receiver is obtained by averaging over all possible interfering symbol combinations. Thus, the BEP of user k can be expressed in the form

$$P_{e,k} = \frac{1}{|\Xi|^{NK-1}} \sum_{\substack{b \in \Xi^{NK} \\ b_k^{(n)}=1}} \Pr[error|b] \qquad (5.31)$$

where $|\Xi|$ denotes the cardinality of the modulation alphabet Ξ, and $\Pr[error|b]$ denotes the probability of error conditioned on the interfering symbols. Expressions for the conditional error probability $\Pr[error|b]$ in Eq. (5.31) are derived in Sections 4.1 and 4.2 for the cases MRC before equalization and equalization before MRC, respectively. Since the averaging over all possible interfering data sequences in Eq. (5.31) is often an overwhelmingly complex operation, approximations are considered in Section 4.3. A few examples are considered in Section 4.4.

4.1   COMBINING BEFORE EQUALIZATION

In AWGN channels with BPSK modulation, a decision error occurs if the decision variable $u_{[MUD]k}^{(n)}$ for user k is less than zero given bit $b_k^{(n)} = 1$ was transmitted. Thus, the conditional error probability can be expressed in the form

$$\Pr[error|b] = Q\left(\frac{\mathrm{Re}(t_k^H \mathcal{H} b)}{\sqrt{\sigma^2 t_k^H t_k}}\right), \qquad (5.32)$$

where t_k is the $(PK+k)$th column of the equalizer matrix $\mathcal{T}$. Since the equalizer impulse response depends on the channel realizations, it is impossible to obtain a general closed form expression for the BEP in a fading channel. The most efficient way to evaluate the BEP in a fading channel is to simulate various fading realizations, compute the equalizer impulse response and matrix $\mathcal{H}$. They can then be substituted into Eq. (5.32) to compute the BEP value for that particular channel realization. Then the BEP values need to be averaged out over numerous channel realizations. In other words, the impact of Rayleigh fading is modeled by computer simulation, whereas the impact of AWGN is treated analytically.

4.2   EQUALIZATION BEFORE COMBINING

The bit error probability of linear multiantenna multiuser receivers is analyzed in this section. In AWGN channels, the analysis can be performed with similar principles as in Section 4.1. Since in this case the equalizer impulse response does not depend on the channel realizations, it is now possible to obtain a general closed form expression for the BEP also in a Rayleigh fading channel. We will concentrate on finding the expression in this section. The analysis is based on the use of the characteristic function of Rayleigh distribution [69, 70]. The method has been earlier applied to the performance analysis of single-antenna multiuser receivers in [65, 66].

The decision variable in Eq. (5.23) can be expressed as

$$\mathbf{g}_k^H \mathbf{v}_k = \boldsymbol{\nu}_k^H \mathbf{Q} \boldsymbol{\nu}_k, \tag{5.33}$$

where

$$\mathbf{Q} = \frac{1}{2} \begin{pmatrix} \mathbf{0}_{ML} & \mathbf{I}_{ML} \\ \mathbf{I}_{ML} & \mathbf{0}_{ML} \end{pmatrix} \in \{0, \frac{1}{2}\}^{2ML \times 2ML}, \tag{5.34}$$

and $\boldsymbol{\nu} = (\mathbf{g}_k^T, \mathbf{v}_k^T)^T \in \mathbb{C}^{2L}$. The linear filter output vector $\mathbf{v}_k | \mathbf{b}, b_k^{(n)} = 1$ conditioned on the data symbol vector $\mathbf{b}, b_k^{(n)} = 1$ is a complex Gaussian random vector. The combining vector $\mathbf{g}_k$ is also complex Gaussian due to the Rayleigh fading assumption. Thus, the conditional probability of bit error for binary phase sift keying (BPSK) modulation of user k can be expressed as [69]

$$\Pr[error|\mathbf{b}] = \sum_{\substack{i=1 \\ \lambda_i < 0}}^{2ML} \prod_{\substack{j=1 \\ j \neq i}}^{2ML} \frac{1}{1 - \frac{\lambda_j}{\lambda_i}}, \tag{5.35}$$

where $\lambda_i, i = 1, 2, \ldots, 2ML$ are the eigenvalues of the matrix $\boldsymbol{\Sigma}_{\boldsymbol{\nu}|\mathbf{b}} \mathbf{Q}$, and

$$\boldsymbol{\Sigma}_{\boldsymbol{\nu}|\mathbf{b}} = \begin{pmatrix} \boldsymbol{\Sigma}_{\mathbf{g}_k} & \boldsymbol{\Sigma}_{\mathbf{g}_k, \mathbf{v}_k | \mathbf{b}} \\ \boldsymbol{\Sigma}_{\mathbf{g}_k, \mathbf{v}_k | \mathbf{b}}^H & \boldsymbol{\Sigma}_{\mathbf{v}_k | \mathbf{b}} \end{pmatrix} \tag{5.36}$$

is the covariance matrix of the vector $\boldsymbol{\nu}$. The kth users' average bit error probability for a linear equalizer at the symbol interval $t \in \left[nT, (n+1)T \right)$ is obtained by averaging over all possible interfering symbol combinations. For BPSK modulation it can be expressed in the form

$$P_k = \frac{1}{2^{NK-1}} \sum_{\substack{\mathbf{b} \in \{-1, 1\}^{NK} \\ b_k^{(n)} = 1}} \Pr[error|\mathbf{b}] \tag{5.37}$$

The analysis method is very practical for numerical performance evaluation. However, it has one disadvantage. If two or more eigenvalues are equal to each other, the method cannot be applied. This is the case in single-user systems, if the propagation paths have equal magnitude. In a multiuser case, the relationship between the eigenvalues and the multipath power profile is not as simple. Due to the crosscorrelations between users' spreading signals the eigenvalues differ usually even if the propagation paths have equal magnitudes. There is also a simple way to avoid the problem in single-user case. The magnitudes of propagation paths can be set to, e.g., 0.51 and 0.49 instead of 0.5 and 0.5. This causes only an unobservable change in the BEP results, but it is enough for the eigenvalues to differ from each other.

The structure of the covariance matrices is further elaborated next. By Eq. (5.29) the covariance of the combiner vector is

$$\Sigma_{g_k} = \Sigma_{v[MAI+N]k}^{-1} \Gamma_k \Sigma_{c_k} \Gamma_k^H \Sigma_{v[MAI+N]k}^{-1}, \qquad (5.38)$$

where Σ_{c_k} is the covariance matrix of the channel gain vector c_k. By Eqs. (5.25) and (5.7)

$$\Sigma_{v_k|b} = W_k^H \left(RB\Sigma_c BR + \sigma^2 R \right) W_k. \qquad (5.39)$$

By Eqs. (5.29), (5.25) and (5.7)

$$\Sigma_{g_k,v_k|b} = \Sigma_{v[MAI+N]k}^{-1} \Gamma_k \Sigma_{c_k,c} B^H R, \qquad (5.40)$$

where $\Sigma_{c_k,c}$ is the covariance matrix between the vector of the kth user's multipath gains and the vector including the multipath gains of all the users.

4.3   APPROXIMATIONS

Two ways to approximate the averaging over the interfering symbol combinations in Eq. (5.31) are briefly addressed here. They a *semianalytic* or *quasianalytic* BEP evaluation , and the other is the *Gaussian approximation* of the interference term at the linear filter output. The most straightforward approximation is to apply a semianalytic evaluation of Eq. (5.31). In other words, only a (small) subset of the possible interfering data symbol vectors $b, b_k^{(n)} = 1$ are generated in a random manner [71]. The values of the conditional error probability $\Pr[error|b]$ are then computed and averaged out over this subset of $b, b_k^{(n)} = 1$. The technique can be viewed as a hybrid of theoretical analysis and Monte-Carlo computer simulations: the contribution of the additive white Gaussian noise, fading, and channel estimation to the BEP is treated via analysis, whereas the impact of the random data pattern is estimated via computer simulation. There is no general rule to determine the number of required averages [71]. It is studied in [65, 66] via numerical examples.

The Gaussian approximation has been applied widely for the evaluation of the BEP of CDMA systems [72, 73]. Usually the performance has been averaged over random spreading sequences, and all possible delay combinations. The Gaussian approximation for the LMMSE receiver without an attempt to average over delays or sequences has been proposed in [74]. There the interference term at the linear equalizer output is approximated by a single Gaussian random variable with zero mean and variance equal to that of the actual interference term. The Gaussian approximation is valid if the individual interference terms have equal variance. This is the case under two possible conditions:

1. there is no near-far problem, i.e., the elements in $\mathcal{A}$ are (approximately) equal, or

2. the receiver filter w_k of user k (or receiver filters $\tilde{w}_{k,l}$, $l = 1, 2, \ldots, L$ in fading channel case) is near-far resistant.

4.4   EXAMPLES

Performance improvements due to multiple antennas is illustrated by numerical examples. BPSK data and spreading modulation with coherent detection are considered. A length 16 random spreading sequences are used, i.e., the processing gain is 16. The channels are assumed to be Rayleigh fading two-path ($L = 2$) channels with equal average energy for the two multipath components[4]. In addition to the multiantenna multiuser results, the bit error probability of the conventional rake receiver with a single-antenna in a single-user system is presented as a reference. The result is referred to as matched filter bound (MFB) in the sequel.

Diversity reception with several antenna elements is considered here. The average signal-to-noise ratios of users are assumed constant in each sensor, i.e., $E_{k,m}/\sigma^2 = E_{k,m'}/\sigma^2, \forall k, m, m'$. Results are presented in Figs. 5.1 and 5.2. The fading processes of the two multipath components are assumed independent on one another in each sensor, i.e., $E(c_{k,l,m}c^*_{k',l',m'}), \forall k, l, m, k', l', m', l \neq l'$. The theoretical Clarke's correlation function [75, Sec. 5.4] for carrier frequency of $f_c = 2$ GHz has been applied. The antenna elements are assumed to be placed on a bar a length of which is fixed to either 10 cm or 50 cm. The increasing number of antenna elements puts the new elements closer and closer each other, and, thus, increases correlation between fading processes. The gains in the SNR requirement at bit error probability of 10^{-2} are summarized in Tables 5.1 and 5.2, when the correlations of the fading processes between sensors are assumed to be zero. The gains of adding a new diver-

---

[4]Actually, the energies were assumed to be 0.51 and 0.49 to avoid the numerical problems in single-user system.

sity antenna element for the rake and LMMSE receivers are included in Table 5.1. The gains of having the LMMSE receiver instead of the rake receiver for different numbers of sensors are presented Table 5.2.

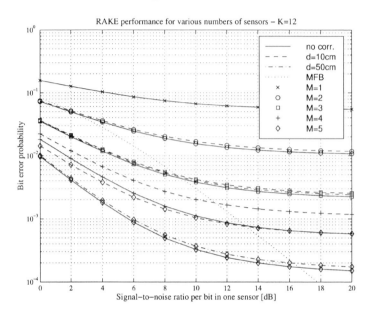

Figure 5.1 Bit probabilities of the rake receiver for various numbers of diversity antennas versus average signal-to-noise ratio; $K = 12, L = 2$.

Table 5.1 SNR gains of adding diversity antennas at BEP of 10^{-2}.

M	$L = 1$		$L = 2$	
	rake	LMMSE	rake	LMMSE
$1 \rightarrow 2$	–	8.5 dB	–	6.5 dB
$2 \rightarrow 3$	–	4 dB	over 15 dB	3 dB
$3 \rightarrow 4$	5 dB	2.3 dB	3 dB	2 dB
$4 \rightarrow 5$	2 dB	1.7 dB	1.5 dB	1 dB

The Clarke's autocorrelation model reveals that the sensor signals can be considered to be essentially uncorrelated, if the antenna elements are placed on a bar of 50 cm. With a bar of 10 cm, there is a significant reduction in the diversity gain with five antennnas, and a noticeable reduction with four

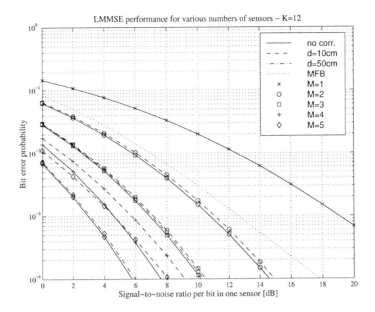

Figure 5.2 Bit probabilities of the LMMSE receiver for various numbers of diversity antennas versus average signal-to-noise ratio; $K = 12$, $L = 2$.

Table 5.2 SNR gains of having LMMSE receiver instead of rake receiver at BEP of 10^{-2}.

M	$L = 1$	$L = 2$
1	over 30 dB	−
2	16 dB	over 30 dB
3	2 dB	4 dB
4	1 dB	2 dB
5	0.5 dB	1 dB

antennas. If a fixed correlation of 0.7 is assumed, the reductions in the diversity advantage are significant with $M \leq 3$. However, adding more antennas gives always a further advantage.

If diversity antennas and multiuser receivers are viewed as competing techniques, it can be seen that three antenna elements are required with rake receiver to outperform a single-antenna LMMSE receiver at low SNR's in a two-path channel. In a single-path channel, two antenna elements in a rake receiver are sufficient to outperform the single-antenna LMMSE receiver. This is understandable, since the single-path channel provides no multipath diversity. If very low bit error probabilities are required, this is not the case; i.e., at high SNR's the rake receiver cannot outperform the LMMSE receiver. It can be seen from Table 5.2 that the gains provided by the LMMSE receiver are larger with less spatial diversity. Gains are larger in a two-path channel than in the one-path channel, since there is intersymbol interference (ISI) and interpath interference (IPI) present in the two-path channel. Both ISI and IPI are also suppressed by the LMMSE equalizer. From Table 5.1 it can be seen that the gain due to the first diversity antenna ($M = 1 \rightarrow 2$) is so large that it is useful to choose $M \geq 2$ even in a two-path channel, if possible. In a single-path channel, it is probably desirable to have at least three antenna elements, if possible. However, it may be more advantageous to implement a multiuser receiver rather than a four-branch antenna diversity. On the other hand, if the use of multiple antennas is out of question[5], the use of a LMMSE receiver yields a significant gain, as seen from Table 5.2. Obviously, the best performance is reached with a combination of antenna diversity and a multiuser receiver.

5   CHIP EQUALIZATION

The basic concepts of linear multiuser equalization have been described in Section 3, but the implementation of the receivers has not been widely considered. While it is not the topic of this paper, it is still an issue which needs to be kept in mind. If the spreading signals of users are periodic with short enough period, the impulse responses of the linear equalizers remains constant over a relatively long times. In a such case it may be possible to implement the matrix inversions needed. Another option is to use some iterative algorithms to implement the equalizers [15, 46]. Probably the simplest way to device an equalizer is to use an adaptive algorithm [56, 57, 58, 59, 60]. However, if the periods of spreading signals are long, as is the case in UMTS W-CDMA downlink, these adaptive solutions become useless. In such a case, a different approach needs to be taken. One promising approach is the topic of this section. First, the system model is briefly redefined. Then based on the redefined system model, the linear equalizers are formulated as chip level channel equalizers.

5.1   SYSTEM MODEL

The downlink DS-CDMA signal differs from the general DS-CDMA signal model presented in Section 2.1 in two aspects. Firstly, the transmitted signals from a sector of a basestation are typically synchronous, and also orthogonal, if orthogonal signature waveforms are employed. Secondly, all signals from

---

[5]For example, in a mobile handset the implementation of multiple antennas may sometimes be impractical. However, the implementation of a linear equalizer might be feasible.

sector p have propagated through the same channel. The complex envelope of the received signal from sector p can be expressed as

$$r_p(t) = \sum_n \sum_{k=1}^{K_p} b_{k,p}^{(n)} A_{k,p} \sum_{l=1}^{L_p} c_{l,p}^{(n)} s_{k,p}(t - nT - \tau_p - \tau_{l,p}) + \eta(t), \quad (5.41)$$

where the notations follow the notations of Eq. (5.1), except here $\tau_p \in (0, T)$ is the propagation delay from sector p.

The system model is redefined to emphasize the special structure of downlink signal and to allow a clear presentation of chip equalizers. The received signal is a combination of signals from $\mathcal{P}$ sectors, although the signal from the sector of interest is usually the dominating one, and the signals from the other sectors may be neglectably weak. The discrete-time model of received signal is written as

$$r = \sum_{p=1}^{\mathcal{P}} D_p C_p \sum_{k=1}^{K_p} A_{k,p} S_{k,p} b_{k,p} + n \in \mathbb{C}^{N_s GN}, \quad (5.42)$$

where N_s is the number of samples per chip, G is the spreading factor, N is the length of observation window in symbols,

$$D_p = \left(d_{1,p}^{(1,1)}, \cdots, d_{L,p}^{(1,1)}, d_{1,p}^{(2,1)}, \cdots, d_{L,p}^{(G,1)} d_{1,p}^{(1,2)}, \cdots, d_{L,p}^{(G,N)} \right) \in \mathbb{R}^{N_s GN \times LGN}$$

is the delay and chip waveform matrix, where column vector $d_{l,p}^{(g,n)} \in \mathbb{R}^{N_s GN}$ contains samples from appropriately delayed chip waveform[6],

$$C_p = \text{diag} \left(c_p^{(1,1)}, \cdots, c_p^{(G,1)}, c_p^{(1,2)}, \cdots, c_p^{(G,N)} \right) \in \mathbb{C}^{LGN \times GN}$$

is the block diagonal channel matrix, where column vector $c_p^{(g,n)} \in \mathbb{C}^L$ contains channel coefficients for L paths, $A_{k,p} = A_{k,p} I_{GN}$ contains the average received amplitudes (square roots of powers),

$$S_{k,p} = \text{diag} \left(s_{k,p}^{(1)}, s_{k,p}^{(2)}, \cdots, s_{k,p}^{(N)} \right) \in \mathbb{C}^{GN \times N},$$

is the spreading sequence matrix, where column vector $s_{k,p}^{(n)} \in \Xi_s^G$, Ξ_s is the chip alphabet, contains the spreading sequence for the kth user's nth symbol from pth sector[7],

$$b_{k,p} = \left(b_{k,p}^{(1)}, b_{k,p}^{(2)}, \cdots, b_{k,p}^{(N)} \right)^T \in \Xi^N,$$

---

[6]In the notation $l = 1, 2, \ldots, L$ refers to the path, $p = 1, 2, \ldots, \mathcal{P}$ to the sector, $g = 1, 2, \ldots, G$ to the chip, and $n = 1, 2, \ldots, N$ to the symbol.

[7]The scrambling sequence is included to the spreading sequence. Scrambling sequences are assumed to be sector specific, i.e., only one scrambling sequence is allocated for each sector.

Figure 5.3   Structure of chip equalizer receiver.

where Ξ is the modulation symbol alphabet, contains the transmitted symbols of kth user from pth sector, and $n \in \mathbb{C}^{N_s GN}$ contains samples from white complex Gaussian noise process with covariance $\sigma_n^2 \mathbf{I}_{N_s GN}$.

5.2   CHIP EQUALIZERS

As mentioned earlier, the received signal at the user terminal is mainly from the sector of interest, and it has propagated through a multipath channel. Due to the non-zero cross-correlations between the spreading sequences with arbitrary time shifts, there is interference between the propagation paths (or rake fingers) in the correlator outputs, which appears multiple access interference. Channel equalization prior to the correlation with the spreading code or matched filtering combines the propagation paths restoring the orthogonality of users to some extent in the case of orthogonal spreading sequences[8]. This translates to the suppression of multiple access interference. Such a receiver is discussed e.g. in [76] and is here referred to as chip equalizer. The receiver consists of a linear equalizer inverting the channel transfer function [77]-[78], followed by a single correlator and a decision device, as depicted in Figure 5.3. Since the received signal is equalized on the chip level, adaptive versions of chip equalizers do not rely on the symbol level cyclostationary, as e.g. [56, 58] do, and they can be applied in systems employing long scrambling codes. Several adaptive versions of chip equalizers have been presented in the literature [79]-[85]. Most of the proposed adaptation schemes are either blind or semiblind. By using of blind or semiblind methods the whole received signal from the sector of interest, i.e., the composite signal of all users assigned to the sector, can be utilized in the adaptation, and the signal-to-noise ratio faced in the adaptation is rather high.

Herein we restrict ourselves to the treatment of optimum equalizers. With the system model defined in Eq. (5.42), the chip equalization basically suppresses the term $D_1 C_1$ for arbitrary sector of interest 1. The zero-forcing, or decorrelating, solution for the chip equalization is given by

$$F = (D_1 C_1) \left((D_1 C_1)^{\mathrm{H}} D_1 C_1 \right)^{-1}. \tag{5.43}$$

---

[8] If non-orthogonal codes are employed, the chip equalization can be used with a fixed MUD.

The first term on the right hand side, DC, performs chip waveform matched filtering and multipath combining, and is followed by the conventional zero-forcing equalizer [77]. The decision variable after despreading for arbitrary user 1 can be written as

$$
\begin{aligned}
y_{[F]} &= S_{1,1}^{H} F^{H} r \\
&= S_{1,1}^{H} \left((D_1 C_1)^{H} D_1 C_1 \right)^{-1} (D_1 C_1)^{H} r.
\end{aligned}
$$
(5.44)

Another well-known equalizer, LMMSE equalizer, can be obtained by solving

$$
L = \arg \min_{L} \mathrm{E} \left[\left| L^{H} r - \sum_{k=1}^{K} A_k S_k b_k \right|^2 \right],
$$
(5.45)

where minimization is carried out elementwise. It should be noted that the chip equalizer does not try to suppress the other users' signals at the chip level, but merely equalizes the channel and restores the orthogonality of users. By estimating the total transmitted signal of all users instead of the chips of a single user, the signal-to-noise ratio faced in the estimation problem is significantly better than by estimating only the signal of the desired user.

It can be shown that for Eq. (5.45) the filter L is given by [86]

$$
L = (D_1 C_1 R_\theta C_1^{H} D_1^{H} + R_w)^{-1} D_1 C_1,
$$
(5.46)

where the covariance matrices R_θ for the desired signal and R_w for noise are defined by

$$
R_\theta = \sum_{k=1}^{K_1} A_{k,1}^2 S_{k,1} S_{k,1}^{H}
$$
(5.47)

$$
R_w = \sum_{p=2}^{P} \sum_{k=1}^{K_p} A_{k,p}^2 D_p C_p S_{k,p} S_{k,p}^{H} C_p^{H} D_p^{H} + \sigma_n^2 I.
$$
(5.48)

The equaliser defined by Eq. (5.46) offers the same performance as the conventional LMMSE receiver presented in [56]-[58] and discussed in Section 3.2, since LMMSE estimator commutes over linear transformations [86], like descrambling and despreading. However, the optimal LMMSE solution depends on the spreading sequences of all users due to the term $S_{k,p} S_{k,p}^{H}$ in Eqs. (5.47) and (5.48). The spreading sequences are not true random sequences, and a priori information of them can be utilized in the equalization. Then the optimal solution changes from chip to chip, i.e., it is non-stationary, and an adaptive version of the receiver will not reach the exact optimal coefficients. An approximation for the LMMSE solution can be obtained by taking time-average of the covariance matrices, or by assuming the consecutive chips to be independent. The approximations of the covariance matrices are given by

$$\tilde{R}_\theta = \sum_{k=1}^{K_1} A_k^2 s^2 \mathbf{I} \tag{5.49}$$

$$\tilde{R}_w = \sum_{p=2}^{P} \sum_{k=1}^{K_p} A_{k,p}^2 s^2 D_p C_p C_p^H D_p^H + \sigma_n^2 \mathbf{I}, \tag{5.50}$$

where s^2 is the square value of chip. The approximation is reasonable for the suboptimal equalizer coefficient set due to the inevitable time averaging in the adaptation algorithm. The resulting equalizer can be written as

$$\tilde{L} = \left(\sum_{p=1}^{P} \sum_{k=1}^{K_p} A_{k,p}^2 s^2 D_p C_p C_p^H D_p^H + \sigma_n^2 \mathbf{I} \right)^{-1} D_1 C_1, \tag{5.51}$$

where the scaling factor $\sum_{k=1}^{K} A_k^2 s^2$ is dropped. The decision variable after the descrambling and correlation with spreading sequence is given by

$$y_{[\tilde{L}]} = S_{1,1}^H \tilde{L}^H r \tag{5.52}$$

$$= (D_{1,1} C_{1,1} S_{1,1})^H \left(\sum_{p=1}^{P} \sum_{k=1}^{K_p} A_{k,p}^2 s^2 D_p C_p C_p^H D_p^H + \sigma_n^2 \mathbf{I} \right)^{-1} r.$$

5.3   EXAMPLES

To illustrate the performance improvements that the chip equalizer receivers can offer, a numerical example is given for a single sector system. Bit error probabilities for the presented chip equalizers were evaluated in a Rayleigh fading frequency-selective channel by applying the semianalytical method discussed in Section 4.1. QPSK modulated data employing root raised cosine pulses with roll-off factor of 0.22 was used with coherent detection. Random cell specific scrambling code and Walsh channelization codes were used. There were 12 users in the system with spreading factor 16. All users had equal transmission powers, and no channel coding was assumed. The channel was a Rayleigh fading two-path channel with equal average energy for the two multipath components. Equalizers were fractionally spaced, i.e., no chip waveform matched filter was used, and four samples per chip were taken.

In the Fig. 5.4, BEP's are presented for the conventional rake receiver with known channel response, as well as for the zero-forcing (ZF) and LMMSE chip equalizer receivers. BEP's were evaluated for the both LMMSE chip equalizers given by Eqs. (5.46) and (5.51), denoted as LMMSE (A) and LMMSE (B) in the figure. Also the theoretical single-user bound [87] for the considered channel is given in the figure.

From the results it is easily seen that as the signal-to-noise ratio increases, the BEP of the rake receiver saturates due to MAI. On the other hand, the BEP's of chip equalizer receivers do not exhibit saturation in the studied range. The LMMSE equalizer receiver shows significant BEP improvement when compared to the conventional rake receiver, whereas the ZF equalizer receiver offers performance improvement only at relatively high SNR. The performance difference between ZF and LMMSE chip equalizer receivers is caused by the noise enhancement typical of ZF equalizers [11]. The performance difference between the exact and approximative LMMSE equalizers defined by Eqs. (5.46) and (5.51) is relatively small, indicating that the non-stationarity of the LMMSE solution does not have a significant effect on the performance of the chip equalizer.

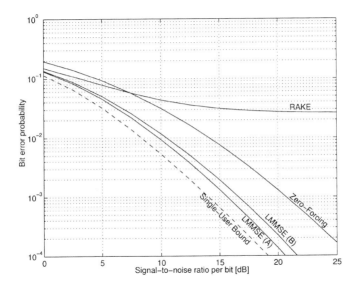

Figure 5.4   Bit error probabilities versus average signal-to-noise ratio for 12 users and spreading factor 16. BEP for the LMMSE receiver given by Eqs. (5.46) (LMMSE (A)) as well as by (5.51) (LMMSE (B)) are presented.

6   CONCLUSIONS

The basic ideas and principles of multiuser receivers were summarized and reviewed. Linear multiuser receiver formulation and combination of MAI suppression with multipath and antenna combining was the main contribution of the chapter. The multiuser equalization can be performed either after maximal ration combining or there can be a separate equalizer in each rake receiver branch. If the latter choice is made, the correlation of noise in the equalizer outputs needs to be taken into consideration when selecting the combining weights. Efficient bit error probability (BEP) evaluation methods for such linear receivers with uncoded transmission were also presented. Thirdly, chip equalizers to equalize the distortions caused by frequency selectivity of the radio channel were described. They are applicable in cases where the signals of the users propagate through a same channel, as is the case of downlink of cellular systems. Their major advantage is that they are applicable also in CDMA systems where the period of spreading sequences is significantly longer than the data symbol period.

Several interesting research items still remain. The main fields in which work is required are the implementation studies and system gain studies. The gains in the presence of practical nonidealities due to implementation (fixed point arithmetic etc.) as well as efficient multiuser receiver architectures need to be found. The other, even more important and difficult, future research area is to translate the achived link level gains to system level gains. That requires introduction of various practical functionalities, like transmit power control and softer handovers in cellular systems. The results of the system level studies needs to provide capacity and coverage gains as well as throughput figures. These are the final performance measures by which the usefulness of the fancy receiver algorithms will be justified.

Acknowledgments

Dr. Matti Latva-aho is acknowledged for useful discussions and fruitful suggestions. Mr. Kimmo Kansanen is acknowledged for numerous comments, which helped to improve the paper.

References

[1] R. L. Peterson, R. E. Ziemer, and D. E. Borth, *Introduction to Spread Spectrum Systems*, Prentice-Hall, Englewood Cliffs, NJ, 1995.

[2] R. C. Dixon, *Spread Spectrum Systems with Commercial Applications*, John Wiley and Sons, New York, NY, 1994.

[3] M. K. Simon, J. K. Omura, R. A. Scholtz, and B. K. Levitt, *Spread Spectrum Communications Handbook*, McGraw-Hill, New York, USA, 1994.

[4] A. J. Viterbi, *CDMA: Principles of Spread Spectrum Communication*, Addison-Wesley Wireless Communications Series. Addison-Wesley, Reading, MA, USA, 1995.

[5] T. Ojanpera and R. Prasad, *Wideband CDMA for Third Generation Mobile Communications*, Artech House, London, UK, 1998.

[6] H. Holma and A. Toskala, Eds., *Wideband CDMA for UMTS*, John Wiley and Sons, New York, 2000.

[7] S. Verdu, "Minimum probability of error for asynchronous Gaussian multiple-access channels," *IEEE Trans. Inform. Th.*, vol. 32, no. 1, pp. 85–96, Jan. 1986.

[8] S. Moshavi, "Multi-user detection for DS-CDMA communications," *IEEE Commun. Mag.*, vol. 34, no. 10, pp. 124–137, Oct. 1996.

[9] M. Juntti and S. Glisic, "Advanced CDMA for wireless communications," in *Wireless Communications: TDMA Versus CDMA*, S. G. Glisic and P. A. Leppanen, Eds., chapter 4, pp. 447–490. Kluwer, 1997.

[10] S. Verdu, *Multiuser Detection*, Cambridge University Press, Cambridge, UK, 1998.

[11] J. G. Proakis, *Digital Communications*, McGraw-Hill, Inc., New York, USA, 3rd edition, 1995.

[12] A. Paulraj and C. Papadias, "Space-time processing for wireless communications," *IEEE Sign. Proc. Mag.*, vol. 14, no. 6, pp. 49–83, Nov. 1997.

[13] R. Kohno, "Spatial and temporal communication theory using adaptive antenna array," *IEEE/ACM Pers. Commun.*, vol. 5, no. 1, pp. 28–35, Jan. 1998.

[14] R. Wichman and A. Hottinen, "New adaptive multiuser detection technique for CDMA mobile receivers," in *Proc. IEEE Int. Symp. Personal, Indoor and Mobile Radio Commun.*, Osaka, Japan, Sept. 12–15 1999.

[15] M. Juntti, *Multiuser Demodulation for DS-CDMA Systems in Fading Channels*, vol. C106 of *Acta Universitatis Ouluensis, Doctoral thesis*, University of Oulu Press, Oulu, Finland, 1997.

[16] M. J. Juntti, "Performance of multiuser detection in multirate CDMA systems," *Wireless Pers. Commun., Kluwer*, vol. 11, no. 3, pp. 293–311, Dec. 1999.

[17] M. J. Juntti and B. Aazhang, "Finite memory-length linear multiuser detection for asynchronous CDMA communications," *IEEE Trans. Commun.*, vol. 45, no. 5, pp. 611–622, May 1997.

[18] S. Kandala, E. S. Sousa, and S. Pasupathy, "Multi-user multi-sensor detectors for CDMA networks," *IEEE Trans. Commun.*, vol. 43, no. 2/3/4, pp. 946–957, Feb./Mar./Apr. 1995.

[19] R. Lupas and S. Verdu, "Linear multiuser detectors for synchronous code-division multiple-access channels," *IEEE Trans. Inform. Th.*, vol. 34, no. 1, pp. 123–136, Jan. 1989.

[20] R. Lupas and S. Verdu, "Near-far resistance of multiuser detectors in asynchronous channels," *IEEE Trans. Commun.*, vol. 38, no. 4, pp. 496–508, Apr. 1990.

[21] Z. Xie, R. T. Short, and C. K. Rushforth, "A family of suboptimum detectors for coherent multiuser communications," *IEEE J. Select. Areas Commun.*, vol. 8, no. 4, pp. 683–690, May 1990.

[22] L. B. Nelson and H. V. Poor, "Iterative multiuser receivers for CDMA channels: An EM-based approach," *IEEE Trans. Commun.*, vol. 44, no. 12, pp. 1700–1710, Dec. 1996.

[23] B.H. Fleury, M. Tschudin, R. Heddergott, D. Dahlhaus, and K. Pedersen, "Channel parameter estimation in mobile radio environments using the SAGE algorithm," *IEEE J. Select. Areas Commun.*, vol. 17, no. 3, pp. 434–450, Mar. 1999.

[24] M. K. Varanasi and B. Aazhang, "Multistage detection in asynchronous code-division multiple-access communications," *IEEE Trans. Commun.*, vol. 38, no. 4, pp. 509–519, Apr. 1990.

[25] R. Kohno, H. Imai, and M. Hatori, "Cancellation technique of co-channel interference in asynchronous spread-spectrum multiple-access systems," *IEICE Trans. Commun.*, vol. 65-A, pp. 416–423, May 1983.

[26] R. Kohno, H. Imai, M. Hatori, and S. Pasupathy, "Combination of an adaptive array antenna and a canceller of interference for direct-sequence spread-spectrum multiple-access system," *IEEE J. Select. Areas Commun.*, vol. 8, no. 4, pp. 675–682, May 1990.

[27] R. Kohno, H. Imai, M. Hatori, and S. Pasupathy, "An adaptive canceller of cochannel interference for spread-spectrum multiple-access communication networks in a power line," *IEEE J. Select. Areas Commun.*, vol. 8, no. 4, pp. 691–699, May 1990.

[28] M. K. Varanasi and B. Aazhang, "Near-optimum detection in synchronous code-division multiple-access systems," *IEEE Trans. Commun.*, vol. 39, no. 5, pp. 725–736, May 1991.

[29] Y. C. Yoon, R. Kohno, and H. Imai, "Cascaded co-channel interference cancellating and diversity combining for spread-spectrum multi-access system over multipath fading channels," *IEICE Trans. Commun.*, vol. E76-B, no. 2, pp. 163–168, Feb. 1993.

[30] U. Fawer and B. Aazhang, "A multiuser receiver for code division multiple access communications over multipath channels," *IEEE Trans. Commun.*, vol. 43, no. 2/3/4, pp. 1556–1565, Feb./Mar./Apr. 1995.

[31] M. Latva-aho and J. Lilleberg, "Parallel interference cancellation in multiuser CDMA channel estimation," *Wireless Pers. Commun., Kluwer*, vol. 7, no. 2/3, pp. 171–195, Aug. 1998.

[32] M. J. Juntti and M. Latva-aho, "Multiuser receivers for CDMA systems in Rayleigh fading channels," *IEEE Trans. Vehic. Tech., to appear*, 2000.

[33] G. H. Golub and C. F. Van Loan, *Matrix Computations, 2nd edn.*, The Johns Hopkins University Press, Baltimore, 1989.

[34] A. J. Viterbi, "Very low rate convolutional codes for maximum theoretical performance of spread-spectrum multiple-access channels," *IEEE J. Select. Areas Commun.*, vol. 8, no. 4, pp. 641–649, May 1990.

[35] P. Patel and J. Holtzman, "Analysis of a simple successive interference cancellation scheme in a DS/CDMA system," *IEEE J. Select. Areas Commun.*, vol. 12, no. 10, pp. 796–807, June 1994.

[36] C. S. Wijting, T. Ojanpera, M. J. Juntti, K. Kansanen, and R. Prasad, "Groupwise serial multiuser detectors for multirate DS-CDMA," in *Proc. IEEE Vehic. Tech. Conf., to appear*, Houston, TX, May 16–20 1999.

[37] T. Ojanpera, *Multirate Multi-User Detectors for Wideband Code Division Multiple Access*, Ph.D. thesis, Technical University of Delft, Delft, The Netherlands, 1999.

[38] K. Kansanen, J. Fan, M. Juntti, and M. Latva-aho, "Groupwise interference cancellation receivers in cellular WCDMA networks," in *Proc. IEEE Vehic. Tech. Conf., to appear*, Tokyo, Japan, May 15–18 2000.

[39] M. K. Varanasi, "Group detection in synchronous Gaussian code-division multiple-access channels," *IEEE Trans. Inform. Th.*, vol. 41, no. 3, pp. 1083–1096, July 1995.

[40] M. K. Varanasi, "Parallel group detection for synchronous CDMA communication over frequency-selective Rayleigh fading channels," *IEEE Trans. Inform. Th.*, vol. 42, no. 1, pp. 116–128, Jan. 1996.

[41] W. Haifeng, J. Lilleberg, and K. Rikkinen, "A new sub-optimal multiuser detection approach for CDMA systems in Rayleigh fading channel," in *Proc. Conf. Inform. Sciences Systems*, The Johns Hopkins University, Baltimore, MD, USA, Mar. 19–21 1997, vol. 1, pp. 276–280.

[42] R. M. Buehrer, A. Kaul, S. Striglis, and B. D. Woerner, "Analysis of DS-CDMA parallel interference cancellation with phase and timing errors," *IEEE J. Select. Areas Commun.*, vol. 14, no. 8, pp. 1522–1535, Oct. 1996.

[43] R. M. Buehrer, N. S. Correal, and B. D. Woerner, "A comparison of multiuser receivers for cellular CDMA," in *Proc. IEEE Glob. Telecommun. Conf.*, London, U.K., Nov. 18–22 1996, vol. 3, pp. 1571–1577.

[44] R. M. Buehrer and B. D. Woerner, "The asymptotic multiuser efficiency of M-stage interference cancellation receivers," in *Proc. IEEE Int. Symp. Personal, Indoor and Mobile Radio Commun.*, Helsinki, Finland, Sept. 1–4 1997, vol. 2, pp. 570–574.

[45] D. Guo, L. K. Rasmussen, S. Sumei, T. J. Lim, and C. Cheah, "MMSE-based linear parallel interference cancellation in CDMA," in *Proc. IEEE Int. Symp. Spread Spectrum Techniques and Applications*, Sun City, South Africa, Sept. 2–4 1998, vol. 3, pp. 917–921.

[46] M. J. Juntti, B. Aazhang, and J. O. Lilleberg, "Iterative implementation of linear multiuser detection for asynchronous CDMA systems," *IEEE Trans. Commun.*, vol. 46, no. 4, pp. 503–508, Apr. 1998.

[47] H. C. Huang, *Combined Multipath Processing, Array Processing, and Multiuser Detection for DS-CDMA Channels*, Ph.D. thesis, Princeton University, Princeton, NJ, USA, Jan. 1996.

[48] M. Latva-aho, *Advanced Receivers for Wideband CDMA Systems*, vol. C125 of *Acta Universitatis Ouluensis, Doctoral thesis*, University of Oulu Press, Oulu, Finland, 1998.

[49] A. Klein and P. W. Baier, "Linear unbiased data estimation in mobile radio systems applying CDMA," *IEEE J. Select. Areas Commun.*, vol. 11, no. 7, pp. 1058–1066, Sept. 1993.

[50] P. Jung and J. Blanz, "Joint detection with coherent receiver antenna diversity in CDMA mobile radio systems," *IEEE Trans. Vehic. Tech.*, vol. 44, no. 1, pp. 76–88, Feb. 1995.

[51] Z. Zvonar, "Combined multiuser detection and diversity reception for wireless CDMA systems," *IEEE Trans. Vehic. Tech.*, vol. 45, no. 1, pp. 205–211, Feb. 1996.

[52] A. Klein, G. K. Kaleh, and P. W. Baier, "Zero forcing and minimum mean-square-error equalization for multiuser detection in code-division multiple access channels," *IEEE Trans. Vehic. Tech.*, vol. 45, no. 2, pp. 276–287, May 1996.

[53] P. Jung, J. Blanz, M. Nasshan, and P. W. Baier, "Simulation of the uplink of JD-CDMA mobile radio systems with coherent receiver antenna diversity," *Wireless Pers. Commun., Kluwer*, vol. 1, no. 2, pp. 61–89, Jan. 1994.

[54] I. N. Psaromiligkos, S. N. Batalama, and D. A. Pados, "On adaptive minimum probability of error linear filter receivers for DS-CDMA channels," *IEEE Trans. Commun.*, vol. 47, no. 7, pp. 1092–1102, July 1999.

[55] D.A. Pados, F.J. Lombardo, and S.N. Batalama, "Auxiliary-vector filters and adaptive steering for DS/CDMA single-user detection," *IEEE Trans. Vehic. Tech.*, vol. 48, no. 6, pp. 1831–1839, Nov. 1999.

[56] P. B. Rapajic and B. S. Vucetic, "Adaptive receiver structures for asynchronous CDMA systems," *IEEE J. Select. Areas Commun.*, vol. 12, no. 4, pp. 685–697, May 1994.

[57] P. B. Rapajic and B. S. Vucetic, "Linear adaptive transmitter-receiver structures for asynchronous CDMA systems," *European Trans. Telecommun.*, vol. 6, no. 1, pp. 21–27, Jan.–Feb. 1995.

[58] U. Madhow and M. L. Honig, "MMSE interference suppression for direct-sequence spread-spectrum CDMA," *IEEE Trans. Commun.*, vol. 42, no. 12, pp. 3178–3188, Dec. 1994.

[59] S. L. Miller, "An adaptive direct-sequence code-division multiple-access receiver for multiuser interference rejection," *IEEE Trans. Commun.*, vol. 43, no. 2/3/4, pp. 1746–1755, Feb./Mar./Apr. 1995.

[60] M. Honig, U. Madhow, and S. Verdu, "Blind adaptive multiuser detection," *IEEE Trans. Inform. Th.*, vol. 41, no. 3, pp. 944–960, July 1995.

[61] M. Latva-aho, "Bit error probability analysis for FRAMES W-CDMA downlink receivers," *IEEE Trans. Vehic. Tech.*, vol. 47, no. 4, pp. 1119–1133, Nov. 1998.

[62] M. Latva-aho, "LMMSE receivers for DS-CDMA systems in frequency-selective fading channels," in *CDMA Techniques for 3rd Generation Mobile Systems*, F. Swarts, P. van Rooyen, I. Oppermann, and M. Lotter, Eds., chapter 13. Kluwer, 1998.

[63] M. Latva-aho and M. Juntti, "LMMSE detection for DS-CDMA systems in fading channels," *IEEE Trans. Commun., to appear*, vol. 48, no. 3, Mar. 2000.

[64] M. Latva-aho and M. Juntti, "Modified LMMSE receiver for DS-CDMA – Part I: Performance analysis and adaptive implementations," in *Proc. IEEE Int. Symp. Spread Spectrum Techniques and Applications*, Sun City, South Africa, Sept. 2–4 1998, vol. 2, pp. 652–657.

[65] M. J. Juntti and M. Latva-aho, "Bit error probability analysis for linear receivers for DS-CDMA systems in fading channels," in *Proc. IEEE Int. Conf. Commun.*, Vancouver, Canada, June 6–10 1999, vol. 1, pp. 51–56.

[66] M. J. Juntti and M. Latva-aho, "Bit error probability analysis of linear receivers for CDMA systems in frequency-selective fading channels," *IEEE Trans. Commun.*, vol. 47, no. 12, Dec. 1999.

[67] M. Juntti, "Performance analysis of linear multiuser receivers for CDMA in fading channels with antenna diversity," in *Proc. IEEE Int. Symp. Personal, Indoor and Mobile Radio Commun.*, Osaka, Japan, Sept. 12–15 1999, vol. 1, pp. 65–69.

[68] M. J. Juntti, "Performance analysis of linear multisensor multiuser receivers for CDMA in fading channels," *IEEE J. Select. Areas Commun.: Wireless Communication Series, preliminarily accepted*, 1999.

[69] M. J. Barrett, "Error probability for optimal and suboptimal quadratic receivers in rapid Rayleigh fading channels," *IEEE J. Select. Areas Commun.*, vol. 5, no. 2, pp. 302–304, Feb. 1987.

[70] V.-P. Kaasila and A. Mammela, "Bit-error probability for an adaptive diversity receiver in a Rayleigh-fading channel," *IEEE Trans. Commun.*, vol. 46, no. 9, pp. 1106–1109, Sept. 1998.

[71] M. C. Jeruchim, P. Balaban, and K. S. Shanmugan, *Simulation of Communication Systems*, Applications of Communications Theory. Plenum Press, 1992.

[72] R. K. Morrow Jr. and J. S. Lehnert, "Bit-to-bit error dependence in slotted DS/SSMA packet systems with random signature sequences," *IEEE Trans. Commun.*, vol. 37, no. 10, pp. 1052–1061, Oct. 1989.

[73] M. O. Sunay and P. J. McLane, "Calculating error probabilities for DS CDMA systems: When not to use the Gaussian approximation," in *Proc. IEEE Glob. Telecommun. Conf.*, London, U.K., Nov. 18–22 1996, vol. 3, pp. 1744–1749.

[74] H. V. Poor and S. Verdu, "Probability of error in MMSE multiuser detection," *IEEE Trans. Inform. Th.*, vol. 43, no. 3, pp. 858–871, May 1997.

[75] J. D. Parsons, *The Mobile Radio Propagation Channel*, Pentech Press, London, 1992.

[76] A. Klein, "Data detection algorithms specially designed for the downlink of CDMA mobile radio systems," in *Proc. IEEE Vehic. Tech. Conf.*, Phoenix, USA, May 4–7 1997, vol. 1, pp. 203–207.

[77] S. Qureshi, "Adaptive equalization," *Proc. IEEE*, vol. 73, no. 9, pp. 1349–1387, Sept. 1985.

[78] D. P. Taylor, G. M. Vitetta, B. D. Hart, and A. Mammela, "Wireless channel equalization," *European Trans. Telecommun.*, vol. 9, no. 2, pp. 117–143, Mar./Apr. 1998.

[79] C. D. Frank and E. Visotsky, "Adaptive interference suppression for direct-sequence CDMA systems with long spreading codes," in *Proc. Annual Allerton Conf. Communication Control and Computing*, Allerton House, Monticello, USA, Sept. 23-25 1998.

[80] I. Ghauri and D. T. M. Slock, "Linear receivers for the DS-CDMA downlink exploiting orthogonality of spreading sequences," in *Proc. 32th Asilomar Conf. on Signals, Systems and Comp.*, Asilomar, CA, Nov. 1–4 1998, vol. 1, pp. 650–654.

[81] S. Werner and J. Lilleberg, "Downlink channel decorrelation in CDMA systems with long codes," in *Proc. IEEE Vehic. Tech. Conf.*, Houston, USA, July 16–19 1999, vol. 2, pp. 1614–1617.

[82] K. Li and H. Liu, "A new blind receiver for downlink DS-CDMA communications," *IEEE Commun. Letters*, vol. 3, no. 7, pp. 193–195, July 1999.

[83] P. M. Grant, S. M. Spangenberg, D. G. M. Cruickshank, S. McLaughlin, and B. Mulgrew, "New adaptive multiuser detection technique for CDMA mobile receivers," in *Proc. IEEE Int. Symp. Personal, Indoor and Mobile Radio Commun.*, Osaka, Japan, Sept. 12–15 1999, vol. 1, pp. 52–54.

[84] P. Komulainen and M. Heikkila, "Adaptive channel equalization based on chip separation for CDMA downlink," in *Proc. IEEE Int. Symp. Personal, Indoor and Mobile Radio Commun.*, Osaka, Japan, Sept. 12–15 1999, vol. 3, pp. 1114–1118.

[85] M. Heikkila, P. Komulainen, and J. Lilleberg, "Interference suppression in CDMA downlink through adaptive channel equalization," in *Proc. IEEE Vehic. Tech. Conf.*, Amsterdam, The Netherlands, Sept. 19–22 1999, vol. 2, pp. 978–982.

[86] S. Kay, *Fundamentals of Statistical Signal Processing: Estimation Theory*, Prentice-Hall, Englewood Cliffs, NJ, USA, 1993.

[87] M. K. Simon and M.-S. Alouini, "A unified approach to the performance analysis of digital communication over generalized fading channels," *Proc. IEEE*, vol. 86, no. 9, pp. 1860–1877, Sept. 1998.

Chapter 6

SOFTWARE-DEFINED RADIO TECHNOLOGIES

Shinichiro Haruyama

Advanced Telecommunication Laboratory, SONY Computer Science Laboratories, Inc.

haruyama@csl.sony.co.jp

Abstract   Thanks to recent advancement of semiconductor technology, it is now possible to process high-speed communication signals in wireless telecommunication system using as much digital technology as possible. This makes the system very flexible and adaptive. Such a technology is called software-defined radio, or simply software radio. This chapter describes software-defined radio technologies and applications.

Keywords: analog-to-digital converter, ASIC, CDMA, digital signal processor, digital-to-analog converter, direct conversion, downconversion, intermediate frequency, LAN, Nyquist sampling theory, OFDM, open-architecture, reconfigurability, software-defined radio, software radio, SpeakEASY, TDMA, undersampling

1   INTRODUCTION

Progress of semiconductor devices in the 1990s made it possible to implement radio equipment using as much digital technology as possible. Even though the progress has been evolutionary, there seems to be a surge of interest in software-defined radio. A special issue on software-defined radio was published in IEEE Communications Magazine [1] in 1995. Since then, there have been numerous activities whose reports were published in conferences such as [2, 3, 4], and [5]. There have also been some special issues on software-defined radio such as [6, 7, 8], and [9]. One early implementation of software-defined radio was SpeakEASY [10], a US military software-defined radio accommodating different modulation methods, frequencies, etc. SpeakEASY demonstrated digital frequency conversion and wideband signal processing, and showed that modular radio elements (modules for the analog elements, A/D converter, and DSP's) could be integrated on an open-architecture bus. This open-architecture approach increases production volume and reduces costs.

Most radio receivers and transmitters today are similar to those used decades ago. They consist of dedicated analog circuits for filtering, tuning and demodulating/modulating a specific type of waveform. To make radio systems more flexible, a software-defined radio is currently being developed for both communication and broadcast applications. A software-defined radio accommodates a variety of receiver/transmitter programs all on a single hardware platform. The programs on the receiver side perform band pass filtering, automatic gain control, frequency translation, low-pass filtering, and demodulation of the desired signal, and similarly on a transmitter side. Maximizing the number of functions handled digitally allows the radio to take advantage of the flexibility of the digital signal processing circuit.

2   APPLICATIONS OF SOFTWARE-DEFINED RADIO

Software-defined radio technology can be applied to all areas of radio communication and broadcasting. Two representative applications of software-defined radio are examined here: cellular phone and wireless LAN.

2.1   CELLULAR PHONE

Figure 6.1 shows the prediction of the number of cellular phones in the world. While the number of first generation analog cellular phones is decreasing, the number of second-generation cellular phones such as GSM is increasing and it will reach 400 million units by the year 2002. The number of third generation CDMA cellular phones will also increase and it will reach 100 million by 2002. Even though older generation systems will eventually be taken

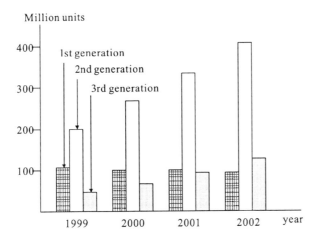

Figure 6.1 Prediction of cellular phone handset units.

over by newer generations, there is a fairly long period when multiple standards co-exist. It will be convenient if the same cellular phone handset or basestation can be used for different services just by changing the software on the system. How realistic this software-defined cellular phone system will become depends on its performance, power consumption, cost, and other business factors, not just convenience. Since software can change the operation of digital hardware, the hardware can be set up in various configuration just by downloading different software. If downloadable software is limited, software-defined radio will not benefit so much from the reconfiguration capability. However, this will become much more feasible economically when there are multitudes of downloadable software, able to accommodate different cellular phone systems.

One of the advantages of software-defined radio is that it can be changed quickly to support multiple standards. For example, in the USA, various cellular standards coexist, and software radio would enable users to overcome the difficulties traveling through areas that use different standards. The software-defined radio handset will reconfigure itself when a user moves from an area employing one cellular standard to another.

Software-defined radio technology can also be applied to cellular network basestations. A new approach to wireless basestation design has the potential of offering significant benefits: reducing the size, complexity, and power consumption of a base station. More importantly, it can support a variety of air interface standards, modulation schemes and protocols simultaneously, and

Table 6.1 Comparison of different cellular phone standards.

	GSM	IS-54/136	PDC	IMT-2000	IS-95
Transmitter Freq. (MHz)	890-915	824-849	940-956/ 1477-1501	1920-1980	1850-1910
Receiver Freq. (MHz)	935-960	869-894	810-826/ 1429-1453	2110-2170	1930-1990
Bandwidth (MHz)	25	25	16/24	60	60
Modulation	GMSK	$\pi/4$ DQPSK	$\pi/4$ DQPSK	BPSK/QPSK	BPSK/QPSK

switching between them whenever required. All the processing is done in software, so it would be possible to load new protocols into the base station as they were developed. In conventional cellular basestations, each channel has a dedicated receiver tuned exclusively to one band. Each of these receivers requires a fair degree of power, size, and expense; it is clear that there can be a lot of expensive receivers in a basestation. Not only are these channels expensive, they are fixed; custom built for a given air interface and modulation standard, and adjusted for a given channel setting. The new approach is to use a single, very high performance wide-band radio receiver to capture and digitize the entire cellular band. Digital mixing and filtering is then used to select and receive individual channels. The high-performance front end of a single radio stage is shared between all channels, instead of the radio per channel of conventional architectures. Signal processing is now all digital, which can be flexibly designed. In addition, the flexibility of the digital stage means that the basestation can be reprogrammed to work with new standards.

Table 6.1 shows the parameters of several cellular phone standards of today and the future. It shows a variety of frequencies used for transmission and reception, different bandwidths, and modulation schemes. In order to make the software-defined radio that will accommodate all of the systems in the table, the following capabilities will be required: RF transmitter and receiver that is able to handle frequencies between 800 MHz and 2000 MHz, a bandwidth of up to 60 MHz, GMSK and QPSK as modulation scheme, and CDMA spread spectrum capability.

2.2   WIRELESS LAN

A popular frequency that has been used for wireless LAN is the 2.4 GHz band, which is called the ISM (Industrial, Scientific, and Medical) band. Regulations regarding the use of the ISM band have not been very strict. As a result, many non-compatible ISM band wireless LAN standards were proposed and developed. On the other hand, new wireless LAN standards are being pro-

Table 6.2   List of proposed wireless LAN standards.

	Bluetooth	HomeRF (SWAP)	IEEE 802.11	BRAN	Wireless 1394/ IEEE 802.11a/ MMAC
Freq (GHz)	2.4	2.4	2.4	5	5
Data rate (Mbps)	1	2	2	54	54
Modulation	FH	FH	FH/DS	OFDM	DMT/OFDM

Figure 6.2   Conventional superheterodyne receiver.

posed as shown in Table 6.2. If people adopt these standards, the incompatibility problem will become less serious than today. However, there will still be a need in the future to be able to handle all the proposed wireless LAN standards. Software-defined radio technology can be applied here too to handle different wireless LAN standards. These standards themselves may be upgraded and improved frequently as new technologies become available. Software-defined radio technology is very suitable for reconfiguring the system to fit the fast-changing wireless LAN standards.

In order to make a software-defined radio for all the wireless LANs in the table, the following capabilities will be required: RF transceiver that is able to handle frequencies between 2.4 GHz and 5 GHz, a data rate up to 54 Mbps, modulation schemes of direct sequence (DS) and frequency hopping (FH) spread spectrum and OFDM.

3   STRUCTURE OF SOFTWARE-DEFINED RADIO

3.1   HOW IS SOFTWARE-DEFINED RADIO DIFFERENT FROM CONVENTIONAL RADIO?

The definition of software-defined radio can easily be understood by comparing it with a conventional radio system. A conventional narrow-band superheterodyne receiver is illustrated in Figure 6.2.

In traditional superheterodyne radio receivers, RF signals from an antenna are received at an antenna and it goes through a band-pass filter. Frequency

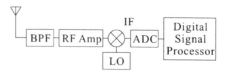

Figure 6.3   IF-sampled software-defined radio.

conversion from the transmitted RF to an Intermediate Frequency (IF) is accomplished by multiplying the RF signal by a sinusoidal Local Oscillator (LO) signal in a mixer. To achieve channel selectivity, additional conversions from higher frequency IFs to lower frequency IFs may also be performed by providing additional mixers and LO signals. An ADC (analog-to-digital converter) then samples the output from the final IF stage, and the digital data is processed by a digital signal processing circuit. Components from an antenna to an ADC are all analog circuits. If more stages of downconversion exist, then more analog components are needed. Analog components have inherent limitations on signal processing capabilities. It is difficult to make a broadband superheterodyne radio receiver, because the analog filters are usually fixed narrowband filters. In addition, analog components are subject to thermal variations and aging effects and also have problems of manufacturing consistency and may require labor-intensive test and alignment. If the number of analog components is reduced, it will result in simplification of the radio system, which in turn will result in higher reliability and reduced cost.

3.2   SOME STRUCTURES OF SOFTWARE-DEFINED RADIO

3.2.1   IF-sampled Software-Defined Radio.
It would be best if all the intermediate analog stages could be replaced by digital components, so that the antenna is directly connected to an ADC. If the received RF signal is in the range of several hundred MHz or above, this is impossible using today's semiconductor ADC technology, which has sampling rates of up to 100 MHz. As a result, realizable software-defined radio today consists of analog components to convert RF signals to IF signals, and ADC and digital components to process the IF signals as shown in Figure 6.3.

A technique called undersampling can be used to sample relatively high frequency IF signals. Nyquist sampling theory requires that a signal to be sampled must be sampled at twice its frequency, to avoid aliasing. If the intermediate frequency f was sampled under the Nyquist sampling rate, it would require a $2f$ sampling rate, which is usually too fast for today's ADC technology. The undersampling of a bandpass filtered signal with a bandwidth of w can be sam-

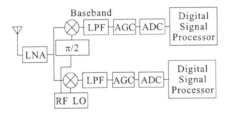

Figure 6.4 Direct conversion software-defined radio.

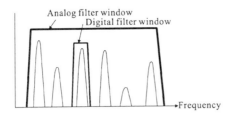

Figure 6.5 Selection of desired signal by a digital filter within an analog filter.

pled at the sampling rate of only $2w$. For example, a CDMA signal with a 6 MHz bandwidth, centered at an IF of 70 MHz, can be captured with a 12-Msps A/D. After undersampling is performed, all the signal components with a frequency above 6 MHz are filtered out. Using this undersampling technique, an ADC with a much slower sampling rate than the IF frequency can be used.

There is an IF technique called near-zero intermediate frequency technology. In the near-zero intermediate frequency technology, the intermediate frequency is very close to DC. If the bandwidth of a signal is B, then the near-zero IF frequency can be as low as B. This analog signal is then converted to a digital signal sampled at the Nyquist sampling rate. The advantage of near-zero intermediate frequency is that DC offset problems do not occur as in the case of direct conversion radio. This will be described in the next section.

3.2.2 Direct Conversion Software-Defined Radio. In direct conversion software-defined radio, RF signals are directly converted to baseband by a quadrature mixer as shown in Figure 6.4. The mixers output in-phase (I) and quadrature-phase (Q) signals, which are then low-pass filtered and gain-controlled before they are digitally sampled.

In the direct conversion software-defined radio, the analog filter passes a broad frequency range, and a desired band within that range can be selected by a digital filter [11] as shown in Figure 6.5. This technique is very useful, for example, when multiple standards using different carrier frequencies and different bandwidths have to be received by one device.

There are some problems that have to be solved for direct conversion receivers. These are the DC offset problem and non-linearity distortion problem. DC offset is a problem DC component from the RF circuit is mixed with direct-converted demodulated signal. Non-linearity distortion is a problem that RF components' non-linearity causes distortion in the demodulated signals. Both of these problems can be adjusted by analog circuits and/or digital processing.

4   KEY COMPONENTS OF SOFTWARE-DEFINED RADIO

Following components are described: MMIC RF component, analog-to-digital converter, and digital signal processing circuit.

4.1   INTEGRATED RF COMPONENT

MMIC (monolithic microwave integrated circuit) technology is used for integrating RF components on one chip. RF components include active components such as transistors and passive components such as resistors, capacitors, and inductors. There are two major materials used for MMIC: GaAs and Si. GaAs is used for frequency ranging from 1 GHz up to 100 GHz, and Si is used for frequency below 10 GHz. The CMOS technology is advancing so that CMOS integrated circuit will be able to handle frequencies of several GHz in a few years. If CMOS analog RF components become available, it will be possible to process not only RF analog signals but also baseband digital signals all on the same chip.

4.2   ANALOG-TO-DIGITAL CONVERTER

Key parameters that define the performance of analog-to-digital convertersare sampling rate and the number of bits per sample. A detailed survey of analog-to-digital converters can be found in [12]. Figure 6.6 shows the relationship between the sampling frequency and the number of bits per sample.

One of the key parameters of an analog-to-digital converter is sampling rate; Software-defined radios sometimes use undersampling as described earlier. When undersampling is done, the sampling rate must be larger than twice the band pass filtered signal bandwidth. Another key parameter is dynamic range. In the conventional approach, each radio only deals with a narrow band; by filtering out interfering signals, the receiver can concentrate on the desired one, adjusting gain to optimize signal-to-noise performance and extracting a weak signal from a noisy background. However, with a wideband receiver, none of the signals should be filtered out, because they are all required. There will be

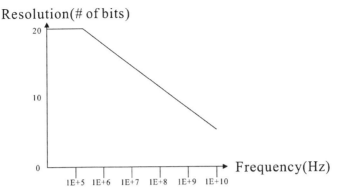

Figure 6.6   Relation between sampling frequency and the number of resolution bits.

a wide range of signals: very strong ones from powerful transmitters nearby, and very weak ones almost buried in noise. As a result, the receiver must have an extremely wide dynamic range for enough sensitivity to accurately recover the weak signals, without their being swamped by the strong ones. It must also be extremely linear; any distortion or harmonics will generate images of strong signals, indistinguishable from true signals.

Performance measure of ADC's can be expressed as $2^m F_s$, where m is the number of bits of a sample and F_s is the sampling range. When the sampling rate is between a few Msps and a few Gsps, which usually covers the software-defined radio applications, this performance measure is usually limited by an aperture jitter. The aperture jitter is the variation of the time difference between the sample command time and the actual time the analog input signal is sampled. The jitter originates from thermal Gaussian distributed noise [13]. The improvement of ADC performance will be done mainly by reducing the aperture jitter, but the progress of sampling bits for a given sampling rate has been fairly slow: only 1.5 bits over the last eight years [12].

There is also an attempt to make a very high speed analog-to-digital converter using superconductor technology [14]. It may be possible to sample analog signals faster than semiconductor analog-to-digital converters. However, there is a problem of the size of a cooling equipment, which is much larger than the ADC device.

4.3   DIGITAL SIGNAL PROCESSING CIRCUIT

When an intermediate frequency signal is sampled by an ADC, signals below IF frequency must be processed digitally as shown in Figure 6.7. The digitized intermediate frequency signal from ADC is down-converted, filtered,

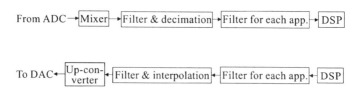

Figure 6.7   Digital processing functions for IF-sampled Software-Defined Radio.

and decimated, before the slower speed signal processing is performed by a DSP. The slower speed signal processing include channel decoding including error correction, and source decoding such as data decompression, description etc. In the transmitter side, the slower signal processing is performed first: source coding such as data compression and encryption, and channel decoding including error correction. The data is then filtered for each application, interpolated, and upconverted before its signal is sent to a DAC.

Signal processing of high speed signals, such as intermediate frequency signals, requires a very high-speed signal processing circuit. The speed may be as high as several thousand MIPS (million instructions per second). Suitable integrated circuits are DSP's (digital signal processors), FPGA (Field Programmable Gate Array), or software-radio-specific ASIC.

A DSP chip does signal processing by fetching instructions and data from memory, does operations, and stores the results back to memory, just like a regular CPU. The difference between a DSP chip and a CPU chip is that a DSP chip usually has a block that does high-speed signal processing, especially a block called MAC (Multiply and Accumulate). By calling different routines in memory, a DSP chip can be reconfigured to perform various functions. Some of commercially available high-speed DSP chips are Texas Instruments' TMS320C6202 and Analog Devices' ADSP-21160M SHARC with the speed of 2000 MIPS and 600 MFLOPS respectively.

ASIC (Application-specific Integrated Circuit) is an integrated circuit that is designed to perform a fixed specific task. Examples of signal-processing-specific ASIC's are DDC (digital down converter) chip, and digital filter chips. The disadvantage of ASIC is that a user cannot change the function of the chip.

FPGA (Field Programmable Gate Array) is able to perform any task by mapping the task to the hardware. On the other hand, FPGA has a reconfigurability capability that ASIC does not have. Reconfigurability is a feature, which enables FPGA to realize any user hardware by changing the configuration data on a chip as many times as needed. Even though the number of gates realizable on one FPGA chip such as Xilinx's Virtex is in the range of 100,000 gates to 1,000,000 gates which is smaller than several million gates of an ASIC, this reconfigurability capability will be very useful in software-defined radio in the future [15, 16]. Typical FPGA's consist of an array of reconfigurable look-up table logic block to implement combinatorial and/or sequential logic, and a reconfigurable routing resource that interconnect logic blocks. Some special signal processing algorithms suitable for FPGA architectures have been developed such as distributed arithmetic algorithm [15, 17, 18]. The distributed arithmetic method uses look-up tables for fast signal processing, which makes LUT-based FPGA's very suitable. The FIR filtering using distributed algorithm, for example, has the same speed whether the number of filter taps is 1 or 100. This makes it suitable for implementing a high-speed filter with large number of taps. Many other applications taking advantage of FPGA architectures will appear in the future. A new FPGA feature that some companies are developing is dynamic reconfiguration. For example, Jbits tool from Xilinx enables users to change configuration of portion of FPGA's while FPGA is operating. This is still a new technology, but this will be a very useful tool when, for example, a receiver needs to reconfigure reception algorithms in order to receive signals that come through a dynamically changing channel.

Software-radio-specific ASIC is a new type of chip that has a fixed portion for common signal processing and a reconfigurable portion that needs to be changed depending on different wireless standards such as different cellular phone standards. Since this is targeted to more specific application than a general-purpose FPGA chip, it is more cost-effective and has a higher performance and consumes less power than FPGA. Some software-radio-specific ASIC's also have dynamic reconfiguration capability.

Among the chips mentioned above, chips that have general-purpose reconfigurability features are DSP's and FPGA's. Table 6.3 shows a table detailing the difference of features between DSP's and FPGA's.

5   STANDARDIZATION

Since the software-defined radio will be used for wireless communication using public radio wave, there have to be a standard regarding radio interface. Even if software-defined radio technology makes any modulation, any carrier frequencies, etc. possible, it should not be allowed to use arbitrary frequencies or modulations in the air. So, standardization or rules about frequency, bandwidth, modulation, the method of download, etc. should be defined before it is used.

In system definition side of software-defined radio, the majority of the functionality of a radio system is achieved by digital hardware with software running on it. When digital hardware/software has a modularity and a hierarchy, then there will be boundaries between modules and between different levels of hierarchy. It will be beneficial if there is a standard interface to interconnect these modules.

Table 6.3   Comparison of FPGA and DSP.

	FPGA chip	DSP chip
Programming Language	VHDL, Verilog	C, Assembly language
Ease of Software Progamming	Fairly easy. However, a programmer needs to understand the hardware architecture before programming.	Easy
Performance	Can be very fast if an appropriate architecture is designed.	Speed is limited by the clock speed of DSP chip.
Reconfigurability	SRAM type FPGAs can be reconfigurable for infinite times.	Can be reconfigurable by changing a program memory content.
Reconfiguration method	Reconfiguration is done by downloading configuration data to a chip electronically	Reconfiguration is done by simply reading program at different memory address.
Areas where FPGAs can outperform DSPs, or vice versa	FIR filter, IIR filter, correlator, convolver, FFT, etc	Signal processing program of sequential nature
Power consumption	Can be minimized if the circuit is designed to save power, or if the power is dynamically controlled	Even if a program A is larger than a program B, power consumption does not change as long as the number of memory chips is the same
Implementation method of MAC	Parallel multiplier/adder or a distributed arithmetic.	Repeated operation of MAC function
Speed of MAC	Can be fast if a parallel algorithm is used. If a filter is implemented using distributed arithmetic, the speed does not depend on the number of taps.	Limited by the speed of MAC operation of a DSP chip. If a filter is implemented, the speed becomes slower if the number of taps increases.
Parallelizm	Can be parallelized to achieve high performance.	DSP chip programming is usually sequential and cannot be parallelized

A group called SDR Forum (Software Defined Radio Forum, http://www.sdrforum.org) has been active since 1996 to propose such a standard interface for software-defined radio. They have been holding several meetings per year, providing input to the International Telecommunication Union (ITU) process for the 3G planning, and publishing technical reports. The most up-to-date report of version number 2.1 was published in November 1999 [19]. Figure 6.8 shows the interface model, which the SDR Forum is proposing.

In the SDR Forum, there is several active groups. Following are some interesting activities of the groups in the Forum. Mobile working group is defining

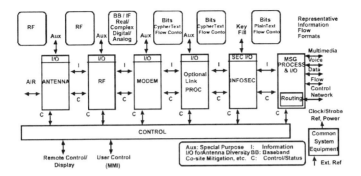

Figure 6.8   SDR Forum Functional Interface Diagram (from SDR Forum technical report 2.1).

interfaces to the SDR services and management structures for SDR control. They are using CORBA (Common Object Request Broker Architecture) and IDL (Interface Definition Language) to define the standard software-defined radio software architecture and planning to finalize one common software radio application for the test of SDR Forum architecture implementations. Basestation working group is defining Use Case description for UML (Unified Modeling Language). Switcher download working group is studying WAP (Wireless Application Protocol) and MExE (ETSI's Mobile Station Application Execution Environment) and planning to contribute to WAP and MexE groups.

It may take a while before this kind of standard will be accepted by the industry. The standardization issue will become extremely important, though, when a high performance software-defined radio platform becomes available to many users. A standard that is not owned by one company or one organization, open architecture, will eventually be accepted in the future.

Acknowledgments

The author wish to thank Dr. Francis Swarts of Sony Computer Science Laboratories, Inc. for his valuable comments.

References

[1] Joseph Mitola III, "Software Radios", IEEE Communications Magazine, pp.24-25, May 1995.

[2] Software Radio Workshop, EU European Commission DGXIII-B, Brussels, Belgium, June 1997.

[3] Software Radio and MMITS Forum, Keio University, April 1, 1998.

[4] First International Software Radio Workshop, ACTS Mobile Communications Summit 1998, Rhodes, Greece, June 8-11, 1998.

[5] Software Defined Radio, IBC UK Conferences Ltd., London, England, October 22, 1998.

[6] Joseph Mitola III, David B. Chester, Shinichiro Haruyama, Thierry Turletti, Walter Tuttlebee, "Globalization of Software Radio", IEEE Communications Magazine, pp.82-123, February 1999.

[7] Joseph Mitola III, Zoran Zvonar, "Topics in Software and DSP in Radio", IEEE Communications Magazine, pp.102-117, August 1999.

[8] Kwang-Cheng Chen, Ramjee Prasad, H. Vincent Poor, "Software Radio", IEEE Personal Communications, pp.12-76, August 1999.

[9] Joseph Mitola III, Zoran Zvonar, "Software and DSP in Radio", IEEE Communications Magazine, pp.68-82, November 1999.

[10] Raymond J. Lackey, Donald W. Upmal, "Speakeasy: The Military Software Radio", IEEE Communications Magazine, Volume 33, No. 5, pp. 55-61, May 1997.

[11] Hitoshi Tsurumi, Yasuo Suzuki, "Broadband RF Stage Architecture for Software-Defined Radio in Handheld terminal Applications", IEEE Communications Magazine, pp.90-95, February 1999.

[12] Robert H. Walden, "Analog-to-Digital Converter Survey and Analysis", the 16th IEEE Instrumentation and Measurement Technology Conference, pp.1558-1562, 1999.

[13] R. J. W. T. Tangelder, H. de Vries, R. Rosing, H. G. Kerkhoff, M. Sachdev, "Jitter and Decision-level Noise Separation in A/D Converters", IEEE Communications Magazine, pp.90-95, February 1999.

[14] Masaaki Katayama, Akira Fujimaki, "Introduction of Superconductive Devices to Software Defined Radio", IEICE SR99-5, pp.31-37, June 1999.

[15] Mark Cummings, Shinichiro Haruyama, "FPGA in the Software Radio", IEEE Communications Magazine, pp.108-112, February 1999.

[16] Chris Dick, Fredric J. Harris, "Configurable Logic for Digital Communications: Some Signal Processing Perspective", IEEE Communications Magazine, pp.107-111, August 1999.

[17] Stanley A. White, "Applications of Distributed Arithmetic to Digital Signal Processing: A Tutorial Review", IEEE ASSP Magazine, pp.4-19, July 1989.

[18] Bernie New, "A Distributed Arithmetic Approach to Designing Scalable DSP Chips", EDN, pp.107-114, August 17, 1995.

[19] Software Defined Radio Forum, Technical Report 2.1, Architecture and Elements of Software Defined Radio Systems as Related to Standards, November 1999.

Chapter 7

SPATIAL AND TEMPORAL COMMUNIATION THEORY BASED ON ADAPTIVE ANTENNA ARRAY

Ryuji Kohno

Div. of Elec. & Comp. Eng., Faculty of Eng., Yokohama National University
79-5 Tokiwadai, Hodogaya, Yokohama 240-8501, JAPAN
kohno@kohnolab.dnj.ynu.ac.jp

Abstract   An adaptive antenna array or a smart antenna can form a desired antenna pattern and adaptively control it if an appropriate set of antenna weights is provided and updated in software. It can be a typical tool for realizing a software radio. An adaptive antenna array can be considered as an adaptive filter in space and time domains for radio communications, so that the communication theory can be generalized from a conventional time domain into both space and time domains. This paper introduces a spatial and temporal communication theory based on an adaptive antenna array, such as spatial and temporal channel modeling, equalization, optimum detection for single user and multiuser CDMA, precoding in transmitter and joint optimization of both transmitter and receiver. Such spatial and temporal processing promises significant improvement of performance against multipath fading in mobile radio communications.

Keywords: Adaptive Array Antenna, Software Antenna, Space-Time Communication Theory, Space-Time Channel Model, Space-Time equalizer, Space-Time Optimum Receiver, Digital Beam Former (DBF), Space-Time Joint Optimum Transmitter and Receiver

1   INTRODUCTION

Recent research interests in the field of wireless personal communications have been moving to the third generation cellular systems for higher quality and variable speed of transmission for multimedia information [1, 2]. For the demand in the third generation wireless personal communications, however, we have several problems which must be addressed. Signal distortion is one of the main problems of wireless personal communications. It can be classified as ISI (Inter-Symbol Interference) due to the signal delay by going through the multipath channel and CCI (Co-Channel Interference) due to the multiple access. There have already been many measures for combatting signal distortion. A traditional equalizer in time domain is useful for short time delay signals [3, 4]. However, when the delay time is large, the complexity of the equalization system increases.

An antenna array, on the other hand, is defined as a group of spatially distributed antennas. The output of the antenna array is obtained by combining properly each antenna output. By this operation, it is possible to extract the desired signal from all received signals, even if the same frequency band is occupied by all signals. An antenna array can reduce the interference according to the arrival angles or directions of arrival (DOA) [5, 6]. Even if the delay time is large, the system complexity does not increase because the antenna array can reduce the interference by using the antenna directivity. Thus, the combination of an antenna array and a traditional equalizer will be able to yield good performance by compensating for drawback each other [7, 8, 9, 10, 11, 12, 13, 14]. It is possible to increase the user capacity, i. e., the number of available users at one base station, by using an antenna array not only in the time domain but also in the space or angular space domain. Therefore, spatial and temporal, i. e. two dimensional signal processing based on an antenna array will become a break-through technique for the third generation of wireless personal communications. This concept has been also successfully used for a long time in many engineering applications such as radar and aerospace technology [15].

Much research for spatial and temporal signal processing using an adaptive antenna array has been pursued in recent years [16, 17, 18, 19, 20, 21]. Research of adaptive algorithms for deriving optimal antenna weights in the time domain such as LMS (Least Mean Squares), RLS (Recursive Least Squares), CMA (Constant Modulus Algorithm) [17] etc, has been proposed from a viewpoint of extending techniques of an adaptive digital filter. On the other hand, there is also research based on DOA estimation from the viewpoint of spectral analysis in the space domain, such as DFT (Discrete Fourier Transfrom) [22], MEM (Maximum Entropy Method) [23], MUSIC [24] and ESPRIT [25, 26]. Adaptive schemes of obtaining the optimal weights are classified into these two groups.

SDMA, i. e. space division multiple access, a new concept of access scheme, is comparable with FDMA, TDMA and CDMA, and can be combined with them for more user capacity. Its research interest is to investigate how much capacity is improved by using an antenna array. Moreover, since communications technology continues its rapid transition from analog to digital and from narrowband to broadband, the fundamental processes, i.e., modulation, equalization, demodulation, etc., have been integrated and implemented in software. This is referred to as *software radio architecture* [27, 28]. Since an adaptive antenna array can form various antenna patterns and adaptively control the pattern with software, it is also named a *smart antenna* or *software antenna*. Thus, the analysis of radio communication systems can be well simulated on a computer. The design of a radio communication system, which includes an air interface, has to consider the combination of each fundamental process. Furthermore, hardware implementation of an adaptive antenna array has been recently reported to ensure performance improvement and to evaluate complexity of implementation [29, 30]. A typical software antenna is a digital beamformer which is implemented by combination of a phased array, down-converter, A/D converter and field programable arrays or digital signal processers [20, 31, 32].

As the above-mentioned trend, the research area for an adaptive antenna array is expanding to many subjects of spatial and temporal signal processing in wireless personal communications. However, there is no communication theory covering the entire subjects based on adaptive antenna arrays. Therefore, the author's group has been researching a spatial and temporal communication theory based on adaptive antenna arrays [33, 34, 35, 36]. This paper briefly introduces an overview of spatial and temporal communication theory for the design and analysis of wireless communication systems using adaptive antenna arrays from a viewpoint of extending a traditional communication theory. I hope this paper will spur further interest in adaptive antenna array and its role in realizing a software radio for wireless personal communications.

2   ADAPTIVE ANTENNA ARRAY

An adaptive antenna array is an antenna array that continuously adjusts its own pattern by means of feedback control. Its comprehensive explanation can be found in many excellent literatures [16, 17, 19, 20]. An adaptive tapped delay line (TDL) antenna array in Fig.7.1, which has digital filter in each antenna element, can also control their own frequency response [10]. The pattern of an array is easily controlled by adjusting the amplitude and phase of the signal from each element before combining the signals.

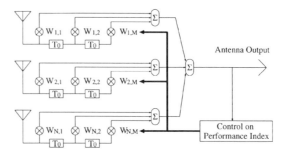

Figure 7.1   An adaptive TDL antenna array with N element antenna and M taps in each element.

When the input signal to the TDL antenna array is $x(t)$, the array output is represented by

$$y(t) = \sum_{n=1}^{N} \sum_{m=1}^{M} x(t - mT_0)w_{n,m} \exp(-jn\varphi), \tag{7.1}$$

where T_0 is the delay between adjacent taps, $w_{n,m}$ is the mth complex tap coefficient of the nth antenna, and N and M are the number of elements and taps at each element antenna respectively. The total number of taps are $N \times M$. φ is the phase difference between the received signal at adjacent antenna elements in a uniform linear array and is given by

$$\varphi = \frac{2\pi S \sin \theta}{\lambda}, \tag{7.2}$$

where λ, S and θ are the wave-length of an incoming signal, the distance between adjacent elements or interelement spacing, and the DOA of the received signal respectively. The antenna transfer function in both spatial frequency or angular space domain, i.e θ and temporal frequency domain, i.e. $f = \omega/2\pi$ is given by

$$H(\omega, \theta) = \sum_{m=1}^{M} \exp(-jm\omega T_0) \sum_{n=1}^{N} w_{n,m} \exp(-jn\varphi) \tag{7.3}$$

This equation (7.3) represents the antenna pattern when ω is a constant, while it represents the frequency response when θ is a constant.

Therefore, the adaptive TDL antenna array can be employed as a tool for signaling, equalization and detection in space and time domains.

3   SPATIAL AND TEMPORAL CHANNEL MODEL

In order to design and analyse an antenna array, a radio transmission model should be modeled in both space and time domains while a traditional communication theory has represented it by a delay profile in time domain. The spatial characteristics, e.g. the angular profile, are important as well as the temporal ones, e.g. the delay profile [37]. The spatial and temporal characteristics of a radio transmission channel are dependent on propagation environments such as indoor, outdoor, various urban and rural areas. A comprehensive discussion on spatial and temporal channel modeling can be found in [20, 38].

For the sake of simplicity, the simple and deterministic model of a multipath channel is employed in this paper to introduce a basis concept of the spatial and temporal communication theory. If time-variation and stochastic properties of delay and angular spread are taken into account, the channel model can be extended to a more practical one. The directional considerations are restircted to the horizontal plane, i.e. azimuth without loss of generality.

A multipath fading channel, such as a mobile radio channel, is modeled in which a transmitted signal from one signal source arrives at the receiver with different angles and delays. The received signal is represented by using two variables, i. e. time t and arrival angle θ.

Each propagation path in a channel is defined by its delay profile or impulse response for a particular DOA θ of the received signal. Thus, the channel can be represented by a spatial and temporal, two dimensional (2D) model like Fig.7.2. Fig.7.3 illustrates such a spatial and temporal or 2D profile of a multipath channel measured by a practical measurement system [38]. From this figure, it is noted that individual propagation paths defined by DOA's have different impulse responses.

Therefore, the impulse response of the kth path $h_k(t)$ with DOA θ_k, ($k = 1, 2, \cdots, K$) is represented by

$$h_k(t) = \sum_{i=1}^{I_k} g_{k,i}\delta(t - \tau_{k,i})\exp(j\psi_{k,i}), \tag{7.4}$$

where $g_{k,i}, \tau_{k,i}$ and $\psi_{k,i}$ denote path amplitude, path delay, and path phase of the ith delayed signal through the kth path, respectively. I_k is the number of delayed signals or the delay spread in the kth path, and $\delta(t)$ is the Dirac delta function.

An equivalent complex baseband representation of the received signal $R_n(t, \theta_k)$ in the nth antenna element is

$$R_n(t, \theta_k) = \sum_{i=1}^{I_k} g_{k,i}S_b(t - \tau_{k,i})$$

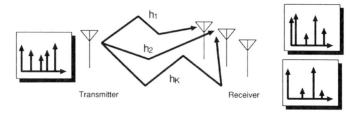

Figure 7.2   Spatial and temporal multipath channel model (2D channel model with paths defined by DOA's and time impulse responses of the paths).

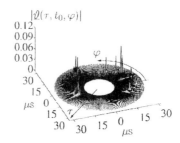

Figure 7.3   A real measured impulse response according to DOA (from Ref [38]).

$$\times \exp\left(-j2\pi nS\frac{\sin\theta_k}{\lambda}\right)\exp(j\phi_{k,i}), \qquad (7.5)$$

where $S_b(t)$ is the complex baseband transmitted signal and $\phi_{k,i}$ is the net phase offset.

4   SPATIAL AND TEMPORAL EQUALIZATION

By using the above-mentioned spatial and temporal channel model, we can derive an extended Nyquist theorem for a known channel [39]. Moreover, for unknown or time-varying channels, various algorithms for updating antenna weights are discussed.

4.1   SPATIAL AND TEMPORAL NYQUIST CRITERION

The Nyquist criterion in space and time domains can be derived from Eqs.(7.1) and (7.5). The array output $y(t)$ can be replaced by $y(t, \Theta)$ because the array output depends on time t and arrival angle set $\Theta = (\theta_1, \theta_2, \cdots, \theta_K)$. The array output is rewritten as

$$y(t, \Theta) = \sum_{k=1}^{K} p(t, \theta_k),\tag{7.6}$$

where $p(t, \theta_k)$ is defined as

$$p(t, \theta_k) = \sum_{n=1}^{N} \sum_{m=1}^{M} \sum_{i=1}^{l_k} w_{n,m} g_{k,i} \exp(j\phi_{k,i})$$
$$\exp(-j2\pi n S \frac{\sin \theta_k}{\lambda}) S_b(t - \tau_{k,i} - mT_0).$$

Suppose that θ_1 represents the desired arrival angle. If $p(t, \theta)$ equals to the symbol s_l at $t = lT_d$ and $\theta = \theta_1$, and equals to zero elsewhere, ISI must be zero. This condition is named the generalized Nyquist criterion in both space and time domains. Then, the criterion is represented by

$$p(lT_d, \theta) = s_l \delta'(t - lT_d, \theta - \theta_1),\tag{7.7}$$

where $\delta'(t, \theta)$ is the two dimensional Dirac delta function and s_l represents the transmitted symbol at $t = lT_d$. This includes the usual Nyquist criterion when $p(t, \theta_k)$ is a function of time only.

4.2   ADAPTIVE S & T EQUALIZATION FOR REDUCING ISI

Several criteria for spatial and temporal equalization such as ZF (Zero Forcing) and MMSE (Minimum Mean Square Error) are available to update the weights and tap coefficients. The ZF criterion satisfies the generalized Nyquist criterion in the noise-free case if there are infinite number of taps and elements. Since the finite number of taps and elements is available in practical noisy multipath channels, there may be some equalization errors in adaptive equalization based on temporal updating algorithms. If the permissible equalization error is given, there may be several combinations of taps and elements which achieve the same equalization error. Therefore, the number of antenna elements can be reduced by increasing the number of taps in some cases, e. g. when the difference in arrival angles is large [39].

Adaptive antenna arrays, such as LMS, RLS, CMA and Applebaum arrays, beamform to track the desired signal and to suppress interfering signals by nulls so as to maximize array output signal-to-noise ratio (SNR) [17, 19, 20]. The Applebaum array is also useful when the DOA of the desired signal is known in advance. The LMS and RLS array doesn't require any knowledge for the DOA of the desired signal, as long as the reference signal correlated with the desired signal can be obtained. However, it is difficult to obtain a reliable reference signal in time-varying channels such as a mobile radio channel. The CMA array can update weights referring a constant envelope of modulated signals but is available only for a constant amplitude modulation in principle. In general, for these temporal updating algorithms, the weights take time to converge to optimum values.

These temporal updating algorithms originate from adaptive digital filters. On the other hand, several algorithms for controlling weights of antenna elements have been derived from the spatial spectrum of spatially sampled signals [22, 23, 24, 25, 26]. The DOA's can be estimated from spatial frequency spectrum, which can be obtained by DFT or MEM for spatially sampled signals. The weight coefficients are updated by the Wiener solution derived from the estimated spatial spectrum. Moreover, the MUSIC algorithm [24] estimates DOA's in noise subspace which is defined by eigen-vectors of covariance matrix of spatially sampled signals, while DFT and MEM do it in signal subspace. MUSIC has better estimation performance than MEM if the noise subspace is larger for uncorrelated signals than signal subspace. These spatial spectral estimation algorithms can be used to obtain optimum or suboptimum weights by using spatial samples at one time instant, i.e. one snap shot. Therefore, if the processing speed is fast enough to track time-variation of channals, these algorithms can be more attractive for a fast fading channel than the temporal updating algorithms.

To combat multipath fading, an adaptive equalizer based on a digital filter in the time domain and a diversity antenna in the space domain have been propsed and investigated in [4, 6, 7, 8, 9, 40]. These are related to spatial and temporal equalization but a diversity antenna is considered as a diversity combiner in the space domain rather than a beamformer.

5   SPATIAL AND TEMPORAL OPTIMUM RECEIVER

In the previous section, spatial and temporal equalization whose purpose is to reduce ISI due to multipath in a channel has been discussed. Viterbi equalization whose purpose is to achieve maximum likelihood sequence estimation utilizing ISI can be also generalized in spatial and temporal domains if an antenna array is employed [11, 12, 14, 41, 42].

In the presence of ISI and AWGN, the tandem structure of a matched filter (MF) and a maximum likelihood sequence estimator (MLSE) or Viterbi Detector (VD) is traditionally considered to be an optimum receiver [43, 44]. The optimum receiver is generalized into space and time domains in this section [34, 45].

5.1   SPATIAL AND TEMPORAL WHITENED MATCHED FILTER

First, the spatially and temporally whitened matched filter (ST-WMF) is derived using a TDL antenna array. The SNR at the TDL antenna array output is represented using the delay operator D as

$$\text{SNR} = \frac{\sigma_s^2 \left| \mathbf{W}^T(D, \Theta) \sum_{k=1}^{K} h_k(D) \mathbf{q}(\theta_k) \right|^2}{\sigma^2 |\mathbf{W}(D, \Theta)|^2}, \qquad (7.8)$$

where $h_k(D)$, $\mathbf{W}(D, \Theta)$, $\mathbf{q}(\theta_k)$, σ_s^2, and σ^2 denote the impulse response of the kth path, the N-dimensional impulse response or weight vector of the array at the arrival angle set Θ, the steering vector for DOA θ_k of the kth path, the variance of the input sequence $x(D)$, and the noise power, respectively. From Schwarz's inequality, the optimal weight vector $\mathbf{W}(D, \Theta)$ (N-dimensional) for maximizing the SNR at the TDL antenna array output is given by the time inversion $h_k(D^{-1})D^{K_0}$ ($k = 1, 2, \cdots, K$) of the impulse response and the directivity information Θ as

$$\mathbf{W}(D, \Theta) = \sum_{k=1}^{K} h_k(D^{-1}) D^{K_0} \mathbf{q}(\theta_k), \qquad (7.9)$$

where $\mathbf{q}^*$ means complex conjugate of $\mathbf{q}$.

$$\mathbf{q}(\theta_k) = \begin{bmatrix} 1 & e^{-j\pi \sin \theta_k} & \cdots & e^{-j(N-1)\pi \sin \theta_k} \end{bmatrix}. \qquad (7.10)$$

where K_0 satisfies $max_{k=1,2,\cdots,K}\{I_k\} \leq K_0$ for the delay spread of kth path I_k.

If a multipath channel is represented by a single time impulse response because an antenna does not distinguish DOA like an omuni-directional antenna, then the time inversion of the channel's impulse response denotes a temporal WMF (T-WMF). If an uniform linear array is used instead of TDL array ($M = 1$), then a spatial WMF (S-WMF) for the kth path is realized by the complex conjugated operation of received phase difference $\exp(jn\pi \sin \theta_k)$ due to DOA θ_k ($n = 1, 2, \cdots, N$). An S-WMF's weight is represented by $\mathbf{W}(\Theta) = [w_1, w_2, \cdots, w_N]^T$, where w_n in the nth antenna element is

$$w_n = \sum_{k=1}^{K} \exp(jn\pi \sin \theta_k). \qquad (7.11)$$

where interelement spacing in array is assumed $\lambda/2$. Therefore, the above $\mathbf{W}(D, \Theta)$ is the generalization of the WMF in both spatial and temporal domains.

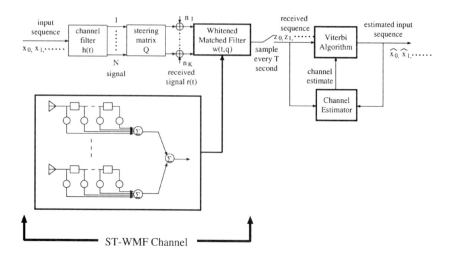

Figure 7.4   Spatial and temporal optimum receiver.

5.2   SPATIAL AND TEMPORAL OPTIMUM RECEIVER

Fig.7.4 shows a VD connected to a ST-WMF which is constructed by a TDL antenna array. We call this a spatial and temporal optimum receiver. As special case, a receiver with a S-WMF & VD and that with a T-WMF & VD are included. The detection algorithm in the proposed receiver is described as follows. (i) each antenna element receives signals, (ii) the received signals in each antenna element are filtered by a ST-WMF, which is matched to the transmission channel impulse response, (iii) the maximum likelihood sequence is estimated from the ST-WMF output. The symbol error probability $P(e)$ of the proposed optimal receiver in the spatial and temporal domains is bounded from above by

$$P(e) \leq \alpha Q_{error}(d_{min}/2\sigma), \tag{7.12}$$

where d_{min} is the minimum Euclidean distance, α is a small constant, and $Q_{error}(\cdot)$ is the error function.

Since ISI is taken into account, the transmission rate $\mathcal{R}_{st}$ is derived as

$$\mathcal{R}_{st} = W \log \frac{\sigma_s^2 \sum_{k=0}^{2K-1} |g_k|^2 + \sigma^2}{\sigma^2}, \tag{7.13}$$

where W and g_k is the signal bandwidth and the delay part of the descrete

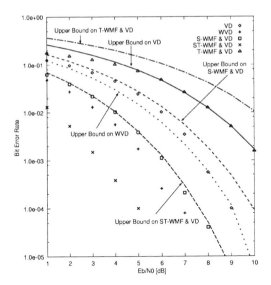

Figure 7.5   BER of optimum receivers in spatial and/or temporal domains.

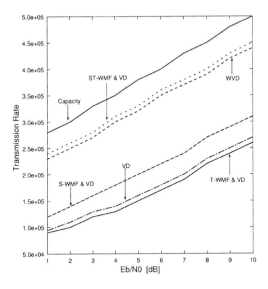

Figure 7.6   Acheivable transmission rate of optimum receivers comparing with Shannon capacity.

channel impulse response of for the kth path, respectively. Figs. 7.5 and 7.6 show the BER and the transmission rate of the proposed ST-WMF and VD receiver in comparison with other receivers. These numerical results are drived in the following case. The number of element antennas and that of taps in each element are $N = 3$ and $M = 3$, respectively. Two incoming signals are assumed, e.g. $h_1(D) = 1.0 + 0.5D$ and $h_2 = 0.7 + 0.2D$ for DOA's $\theta_1 = 30$ (deg) and $\theta_2 = 60$ (deg), respectively.

5.3   SPATIAL AND TEMPORAL OPTIMUM MULTIUSER RECEIVER FOR CDMA

A direct-sequence CDMA (DS/CDMA) mobile radio communication chan-nel is modeled as a channel with both ISI due to multipath and co-channel interference (CCI) due to the correlation between spreading sequences of mul-tiple access users. The optimum multiuser receiver for DS/CDMA detects ev-ery user's data in a sense of MLSE by utilizing CCI as redundant information which multiple access users share. By using an adaptive TDL antenna array, we derive a spatial and temporal optimum multiuser receiver such that MLSE for every user's data can be achieved with both CCI and ISI in present [46]. From a different viewpoint, it is considered that conventional multiuser re-ceivers are designed only in a time domain [47, 48] but it can be generalized into a spatially and temporally optimized multiuser receiver.

The receiver has an extended structure in Fig. 7.4 so that correlators for every user are located in front of the ST-WMF, the ST-WMF is modified to be multiple input/output structure with cross coupling and is followed by multiple VD. The detection algorithm in the proposed receiver is described as follows. (i) Each antenna element receives signals. (ii) Received signals in each element are filtered by each user's correlator. (iii) Each user's correlator output vector is filtered by each user's ST-WMF, which is matched to each user's channel impulse response. (iv) Each user's maximum likelihood sequence is estimated for each user's ST-WMF output, where the path metric is calculated taking into consideration the influence of CCI. Fig. 7.7 shows the BER of the ST optimum multiuser receiver accordiing to different processing gains or spreading ratios of 31, 63 and 127(typed in the figues) in the case of three users; the first user, $h_{1,1} = 0.8 + 0.7D$ for $\theta_{1,1} = 20$ (deg), $h_{1,2} = 0.5D^2$ for $\theta_{1,2} = -45$ (deg), the second user, $h_{2,1} = 3.50.7D$ for $\theta_{2,1} = 15$ (deg), $h_{2,2} = 7.0 + 2.8D^2$ for $\theta_{2,2} = -60$ (deg), the third user, $h_{3,1} = 7.0 + 5.6D$ for $\theta_{3,1} = 25$ (deg), $h_{3,2} = 4.2D^2$ for $\theta_{3,2} = 10$ (deg).

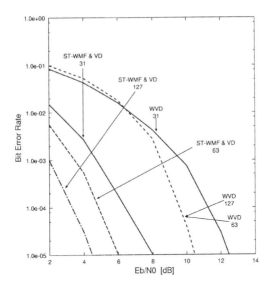

Figure 7.7   BER of optimum CDMA multiuser receiver in spatial and temporal domains according to different processing gains(No. of users=3).

6   SPATIAL AND TEMPORAL JOINT EQUALIZER IN TRANSMITTER AND RECEIVER

6.1   SPATIAL AND TEMPORAL PARTIAL RESPONSE SIGNALING

In the previous sections, we have discussed equalization and detection in a receiver that is useful for unknown or time-varying channels because channel characteristics can be adaptively estimated in a receiver. Equalization in a transmitter that is partial response signaling (PRS) or precoding is also possible in a two-way interactive communication such as time or frequency division duplex (TDD or FDD), because the channel characteristics estimated by received signals can be employed for PRS. If transmitting TDL antenna arrays are prepared in a tranmitter, the spatial and temporal PRS can be carried out [33].

Figure 7.8 shows the transmitting TDL antenna array which consists of J element antennas and M taps in each element. The number of antenna weight sets is L and these antenna weight sets are used for signal transmission in L directions. The transmitting TDL antenna array in a time domain is characterized by the matrix of finite impulse response $\mathbf{W}_{t,l}(D)$ ($l=1,2,\cdots,L$) ($J \times M$-dimensional), where TDL with M taps are used as a precoder or pre-equalizer in a time domain. Although L sets of transmitting TDL antenna

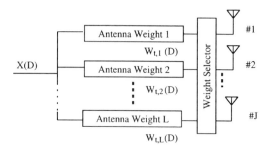

Figure 7.8 Structure of transmitting TDL antenna array (J elements and L sets of antenna weights).

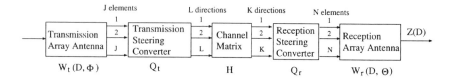

Figure 7.9 Channel including transmitter and receiver.

arrays are required in order to precode data signals for L directions of transmission (DOT's), for the sake of reducing hardware complexity a set of J element antennas can be used by periodically switching L sets of antenna weights like Fig. 7.8. The procedure in the proposed precoder is described as follows. (i) Signals are precoded by temporal PRS for each DOT. (ii) The precoder outputs are appropriately weighted for each DOT. (iii) The weighted signals are transmitted from every element.

Transmitted signals with the direction of transmission (DOT) ϕ_l, ($l=1,2,\cdots,L$) is propagated through a multipath channel and are received with DOA θ_k, ($k=1,2,\cdots,K$). The channel can be represented as an L inputs and K outputs multidimensional channel with cross-coupling. The channel model is represented by Fig. 7.9.

If the channel characteristics are known in a transmitter, the precoder can equalize the channel distortion. However, since the channel has unknown, nonlinear, time-varying factors, a receiving antenna array is required to compensate for the residual distortion.

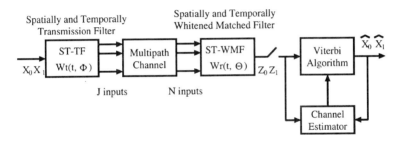

Figure 7.10　Spatial and temporal joint transmitter-receiver system.

6.2　SPATIAL AND TEMPORAL JOINT TRANSMITTER-RECEIVER SYSTEM

If transimitting and receiving TDL antenna arrays are used, then they should be jointly optimized with a certain criterion, e.g. MLSE [49].

Figure 7.10 shows a spatial and temporal joint transmitter- receiver system which consists of ST-transmission filter (ST-TF) based on transmitting TDL array, ST-WMF based on receiving TDL array and Viterbi detector (VD) for MLSE. The detection algorithm for received sequence is described as follows.

1. Each antenna element receives signals.

2. Received signals in each antenna element are filtered by a ST-WMF.

3. The maximum likelihood sequence is estimated for the ST-WMF output by VD.

ST-TF is optimized such that minimum Euclidian distance in a trellis diagram of VD can be maximized. ST-WMF is matched to the impulse response of both ST-TF and the multipath channel.

The symbol error probability $P_{TR}(e)$ and the achievable transmission rate $\mathcal{R}_{st-TR}$ of the proposed joint transmitter-receiver system in the spatial and temporal domains are derived in a similar manner to ST optimum receiver, and are illustrated in Figs.7.11 and 7.12, respectively. These numerical results are drived in the following case. The number of element antennas and that of taps in each element are $J = N = 3$ and $M = 3$, respectively. Two DOT's and two DOA's are assumed, e.g. for the first DOT $\phi_1 = 10$ (deg), $h_1(D) = 1.0 + 0.5D$ and $h_2 = 0.7 + 0.2D$ for DOA's $\theta_1 = 30$ (deg) and $\theta_2 = 60$ (deg), respectively. for the second DOT $\phi_1 = 50$ (deg), $h_1(D) = 0.1 - 0.2D$ and $h_2 = -0.3 + 0.4D$ for DOA's $\theta_1 = 30$ (deg) and $\theta_2 = 60$ (deg), respectively.

Optimum solution of combination between transmitting and receiving TDL antenna arrays, $\mathbf{W_t}$ and $\mathbf{W_r}$ is not unique in a sense of minimum BER at the

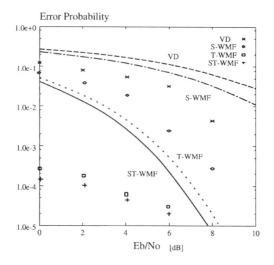

Figure 7.11   BER of spatial and/or temporal joint transmitter-receiver systems.

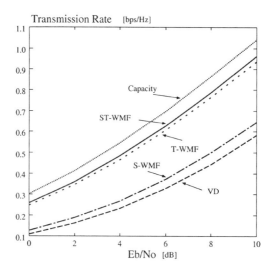

Figure 7.12   Acheivable transmission rate of spatial and/or temporal joint transmitter-receiver systems.

whole output, but there are several combinations in a spatial and temporal joint transmitter-receiver system of Figure 7.10. If balance of hardware complexity between transmitter and receiver is considered, then the complexity of receiving and transmitting antennas in a mobile station (MS) can be minimized by installing a complex transmitting and receiving antenna arrays in a base station (BS) for downlink (from MS to BS) and uplink (from BS to MS), respectively, in a cellular mobile communication system.

6.3   S & T JOINT MULTIUSER TRANSMITTER-RECEIVER SYSTEM FOR CDMA

The spatial and temporal joint transmitter-receiver system can be extended to the multiuser environment in CDMA by the same manner as the spatial and temporal optimum receiver [50]. Fig. 7.13 illustrates that the bit error rate (BER) depends on the number of users. For the limit of pages, specification in this calculation is ommited. The number of element antennas and that of taps in each element are $J = N = 3$ and $M = 3$, respectively, for all users. Fig. 7.14 illustrates that the achievable transmission rate of the proposed system can be close to the channel capacity when the number of users is small.

Although hardware complexy inceases according to the number of accessing users, these figures theoretically prove that the spatial and temporal joint optimization of transmitting and receiving antenna arrays can improve user capacity of CDMA drastic. At BS in a cellular CDMA system, the single adaptive TDL antenna array can be shared for all users' detection if correlators for the users are installed at each element antenna in parallel. Schemes of reducing complexity maintaining capacity improvement should be further studied.

7   CONCLUDING REMARKS

Feasibility of implementing an adaptive antenna array is increasing in higher frequency bands such as millimeter wave band [31, 32]. When an adaptive array antenna is available in practice, the spatial and temporal communication theory will become more important for achieving high-speed and highly reliable radio communications. Moreover, since an adaptive antenna array can be such an almighty antenna that any antenna pattern can be designed with software, it will be a vital tool to carry out a software radio transceiver.

Some idea conditions have been assumed in channel modeling and a physical structure of adaptive antenna array such that readers can easily understand a concept of the spatial and temporal communication theory based on an adaptive antenna array. Analysis and optimization of the systems should be achieved for more practical mobile radio channels.

An adaptive antenna array brings us a new researching paradigm. Not only the inroduced theory but also further researching subjects, such as spa-

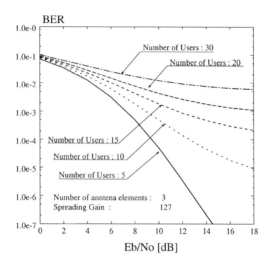

Figure 7.13   BER according to the number of users in S & T joint transmitter-receiver system for CDMA.

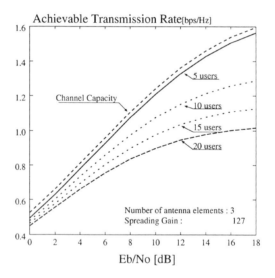

Figure 7.14   Achievable transmission rate according to the number of users in S & T joint transmitter-receiver system for CDMA.

tial and temporal coding [51], modulation, and an adaptive algorithm for a time-varying channel should be also taken into account.

References

[1] D. L. Schilling, *"Wireless Communications Going Into the 21st Century,"* IEEE Trans. Veh. Technol., vol. 43, no. 3, pp. 645 – 652, Aug. 1994.

[2] R. Kohno, R. Meidan, and L. B. Milstein, *"Spread Spectrum Access Methods for Wireless Communications,"* IEEE Commun. Mag., vol. 33, no. 1, pp. 57 – 67, Jan. 1995.

[3] S.U.H.Qureshi, *"Adaptive Equalization,"* Proceedings of the IEEE, vol. 73, no. 9, pp. 1349–1387, Sep. 1985

[4] P. Monsen, *"MMSE Equalization of Interfernce on Fading Diversity Channels,"* IEEE Trans. Commum., vol. com-32, no. 1, pp. 5-12, Jan. 1984.

[5] J. H. Winters, *"Signal Acquisition and Tracking with Adaptive Arrays in Digital Mobile Radio System IS-54 with Flat Fading,"* IEEE Trans. Veh. Technol., vol. 42, no. 3, pp. 377-384, July 1993.

[6] J. W. Mondestino and V. M. Eyuboglu, *"Integrated Multielement Receiver Structures for Spatially Distributed Interference Channels,"* IEEE Trans. Inform. Theory, vol. IT-32, no. 2, pp. 195-219, March 1986.

[7] R. Kohno, H. Imai, M. Hatori and S. Pasupathy, *"Combination of an Adaptive Array Antenna and a Canceller of Interference for Direct-Sequence Spread-Spectrum Mutliple-Acess System,"* IEEE J. Select. Areas Commun., vol. 8, no. 4, pp. 675 – 682, May 1990.

[8] N. Kuroiwa, R. Kohno, H. Imai, *"Design of a Directional Diversity Receiver Using an Adaptive Array Antenna,"* Trans. IEICE Japan, J73-B-II, vol. J73-B-II, no. 11, pp. 755–763, Nov. 1990

[9] R. Kohno, H. Wang, H. Imai, *"Adaptive Array Antenna Combined with Tapped Delay Line Using Processing Gain for Spread-Spectrum CDMA Systems,"* PIMRC'92, pp. 634 – 638, Boston, Massachusetts 1992

[10] R. Kohno, *"Spatial and Temporal Filtering for Co-Channel Interference in CDMA,"* Chapter 3 in Code Division Multiple Access Communications (S.G. Glisic and P.A. Leppanen Ed.) Kluwer Academic Publishers, 1995.

[11] G. E. Bottomley and K. Jamal, *"Adaptive Arrays and MLSE Equalization,"* IEEE Veh. Tech. Conf., pp. 50-54, July 1995.

[12] Y. Doi, T. Ohgane and E. Ogawa, *"ISI and CCI Canceller Combining the Adaptive Array Antennas and the Viterbi Equalizer in a Digital Mobile Radio,"* IEEE Veh. Tech. Conf., pp. 81-85, April 1996.

[13] B. C. Ng, J. T. Chen and A. Paulraj, *"Space-Time Processing for Fast Fading Channels with Co-Channel Interference,"* IEEE Veh. Tech. Conf., pp. 1491 – 1495, April 1996.

[14] S. N. Diggavi and A. Paulraj, *"Performance of Multisensor Adaptive MLSE in Fading Channels,"* IEEE Veh. Tech. Conf., pp. 2148 – 2152, May 1997.

[15] S. Haykin and A. Steinhardt Edited, *"Adaptive Radar Detection and Estimation,"* John Willey & Sons, Inc. 1992.

[16] R. A. Monzingo and W. T. Miller, *"Introdution to Adaptive Arrays,"* John Willey & Sons, Inc. 1980.

[17] R. T. Compton. Jr., *"Adaptive Antennas: Concepts and Performance,"* Prentice Hall, 1988

[18] B. D. V. Veen and K. M. Buckley, *"Beamforming: A Versatile Approach to Spatial Filtering,"* IEEE ASSP Mag., pp. 4 – 24, Apr. 1988.

[19] D. H. Johnson and D.E. Dudgeon., *"Array Siganl Proceesing: Concepts and Techniques,"* Prentice Hall, Inc. 1993.

[20] J. Litva and T. K-Y. Lo, *"Digital Beamforming in Wireless Communications,"* Artech House 1996.

[21] J. S. Thompson, P. M. Grant, and B. Mulgrew, *"Smart Antenna Arrays for CDMA Systems,"* IEEE Pesonal Commun. Mag., vol. 3, no. 5, pp. 16 –25, Oct. 1996.

[22] R. Kohno, C. Yim and H. Imai, *"Array Antenna Beamforming Based on Estimation on Arrival Angles Using DFT on Spatial Domain,"* the 2nd International Symposium on Personal, Indoor and Mobile Radio Communications (PIMRC'91), London, UK, pp. 38 – 43, Sept. 1991.

[23] M. Nagatsuka, N. Ishii, R. Kohno and H. Imai, *"Array Antenna Based on Spatial Spectrum Estimation Using Maximum Entropy Method,"* IEICE Trans. Commun., vol. E77-B, no.5, pp. 624 – 633, May 1994.

[24] R. O. Schmidt, *"Multiple Emitter Location and Signal Parameter Estimation,"* IEEE Trans. Antennas Propagat., vol. AP-34, no.3, pp. 276-280, March 1986.

[25] R. Roy and T. Kailath, *"ESPRIT-Estimation of Signal Parameters via Rotational Invariance Techniques,"* IEEE Trans. Accoust., Speech Signal Process., vol. ASSP-37, pp. 984-995, July 1989.

[26] M. Haardt, *"Efficient One-, Two-, and Multidimensional High-Resolution Array Signal Processing,"* a doctoral thesis of Technical University of Munich, Aachen, Shaker Verlag 1997.

[27] J. Mitola, *"The Software Radio Architecture,"* IEEE Commun. Mag., vol. 33, no. 5, pp. 26 – 38, May 1995.

[28] J. Kennedy and M. C. Sulivan, *"Direction Finding and "Smart Antennas" Using Software Radio Architectures,"* IEEE Commun. Mag., vol. 33, no. 5, pp. 62 – 68, May 1995.

[29] R. Miura, T. Tanaka, I. Chiba, A. Horie and Y. Karasawa, *"Beamforming Experiment with a DBF Multibeam Antenna in a Mobile Satellite Environment,"* IEEE Trans. Antennas Propagat., vol. AP-45, no. 4, pp. 707 – 714, April 1997.

[30] T. Tanaka, R. Miura and Y. Karasawa, *"Implementation of a Digital Signal Processor in a DBF Self-Beam-Steering Array Antenna,"* IEICE Trans. Commun., vol. E80-B, No. 1, pp. 166 – 175, August 1995.

[31] Y. Karasawa and H. Inomata, *"Research on Digital and Optical Beamforming Antennas in Japan,"* Proc. JINA'96, pp. 159 – 168, Nov. 1996.

[32] P.E. Mogensen, et al., *"A Hardware Testbed for Evaluation of Adaptive Antennas in GSM/UMTS,"* Proc. IEEE PIMRC'96, pp. 540 – 544, Oct. 1996.

[33] R. Kohno, *"Information Theoretical Aspect of Adaptive Array Antenna Systems,"* 1995 IEEE International Workshop on Information Theory (ITW'95), Session 2.6, June 1995.

[34] R. Kohno, *"Spatial and Temporal Precoding and Equalization Using Adaptive Array Antenna for Mobile Radio Communications,"* 1995 Allerton Conference on Communication, Control, and Computing, pp. 776–785, Oct. 1995.

[35] N. Ishii, *"Signal Design and Detection Theory Based on an Adaptive Array Antenna in Spatial and Temporal Domains,"* a doctoral thesis in Yokohama National University, Dec. 1996.

[36] R. Kohno, *"Spatial and Temporal Communication Theory Based on Adaptive Array Antenna for Mobile Radio Communications,"* in Part 3 of *Wireless Communications, TDMA versus CDMA,* edited by S. G. Glisic and P. A. Leppanen, Kluwer Academic Publishers, pp. 293 – 321, 1997.

[37] P. C. Eggers, *"TSUNAMI: Spatial Radio Spreading as Seen by Directive Antennas,"* COST 231 TD(94)119, Darmstadt, 1994.

[38] J. J. Blanz, P. Jung and P. W. Baier, *"A Flexibly Configurable Statistical Channel Model for Mobile Radio Systems with Directional Diversity,"* AGARD SPP Symposium, Athenes, Greece, pp. 38-1 – 38-11, 1995.

[39] N. Ishii and R. Kohno, *"Spatial and Temporal Equalization Based on an Adaptive Tapped Delay Line Array Antenna,"* IEICE Trans. Commun., vol. E78-B, no.8, pp. 1162 – 1169, August 1995.

[40] S. Y. Miller and S. C. Schwartz, *"Integrated Spatial-Temporal Detectors for Asynchronous Gaussian Multiple-Access Channels,"* IEEE Trans. on Commun., vol. 43, no. 2/3/4, pp. 396 – 411, Feb./March/April 1995.

[41] R. Krenz and K. Wesolowski, *"Comparison of Several Space Diversity Techniques for MLSE Receivers in Mobile Communications,"* IEEE In-

ternational Symposium on Personal Indoor and Mobile Radio Communications (PIMRC'94), vol. II, pp. 740–744, Sept. 1994.

[42] N. Ishii and R. Kohno, *"Tap Selectable Viterbi Equalizer Combined with Diversity Antenna,"* IEICE Trans. Commun., vol. E78-B, no. 11, pp. 1498 – 1506, Nov. 1995.

[43] G. D. Forney, Jr., *"Maximum-likelihood Sequence Estimation of Digital Sequences in The Presence of Intersymbol Interference,"* IEEE Trans. Inform. Theory, vol. IT-18, no. 5, pp. 363–378, May 1972.

[44] F. R. Magee, jr., J. G. Proakis, *"Adaptive Maximum-Likelihood Sequence Estimation for Digital Signaling in the Presence of Intersymbol Interference,"* IEEE Trans. Inform. Theory, vol. IT-19, no. 1, pp. 120 – 124, Jan. 1973.

[45] M. Nagatsuka, R. Kohno and H. Imai, *"Optimal Receiver in Spatial and Temporal Domains Using Array Antenna,"* ISITA 1994, pp. 893 – 898, Nov. 1994.

[46] M. Nagatsuka and R. Kohno, *"A Spatially and Temporally Optimal Multiuser Receiver Using an Array Antenna for DS/CDMA,"* IEICE Trans. Commun., vol. E78-B, no. 11, pp. 1489 – 1497, Nov. 1995.

[47] R. Kohno , H. Imai and M. Hatori, *"Cancellation techniques of cochannel intererence in asynchronous spread spectrum multiple access systems,"* Trans. IEICE, vol. J66-A, no. 5, pp. 416 – 423, May 1983.

[48] S. Verdu, *"Minimum probability of error for asynchronous Gaussian multiple access channels,"* IEEE Trans. Inform. Theory, vol. IT-32, no. 1, pp. 85 – 96, Jan. 1986.

[49] N. Ishii and R. Kohno, *"Spatial and Temporal Joint Transmitter-Receiver Using an Adaptive Array Antenna,"* IEICE Trans. Commun., vol. E79-B, no. 3, pp. 361-367, March 1996.

[50] N. Ishii and R. Kohno, *"Joint Optimization of Spatial and Temporal Multiuser Equalization in Both Transmitter and Receiver Using an Adaptive Array Antennas for DS/CDMA,"* IEEE GLOBECOM'96, Commun. Theory Mini-Conference, pp. 137 – 141, Nov. 1996.

[51] A. Saifuddin, R. Kohno and H. Imai, *"Integrated Receiver Structure of Staged Decoder and CCI Canceller for CDMA with Multilevel Coded Modulation,"* Europ. Trans. on Telecomm. and Related Technol., vol. 6, no. 1, pp. 9 – 19, Jan.-Feb., 1995.

II

TRENDS IN NEW WIRELESS MULTIMEDIA COMMUNICATION SYSTEMS

Chapter 8

INTELLIGENT TRANSPORT SYSTEMS

Masayuki Fujise
Communications Research Laboratory, Ministry of Post and Telecommunications, Japan
fujise@crl.go.jp

Akihito Kato
Communications Research Laboratory, Ministry of Post and Telecommunications, Japan
akihito@crl.go.jp

Katsuyoshi Sato
Communications Research Laboratory, Ministry of Post and Telecommunications, Japan
satox@crl.go.jp

Hiroshi Harada
Communications Research Laboratory, Ministry of Post and Telecommunications, Japan
harada@crl.go.jp

Abstract   The Communications Research Laboratory (CRL), Ministry of Posts and Telecommunications, Japan has been conducting research and development on Inter-Vehicle Communications (IVC), Radio on Fiber (ROF), Road-Vehicle Communications (RVC), and software radio technologies for Intelligent Transport Systems (ITS) utilizing microwave and millimeter wave. In this chapter, research and development activities for these technologies at CRL are presented.

Keywords: ITS, Radio on Fiber, Road-Vehicle communication, Inter-Vehicle communication, Software Radio, microwave, millimeter wave

1   INTRODUCTION

The development of the transportation in recent years has also held negative sides such as the traffic accident and environmental pollution. ITS (Intelligent Transport Systems) is expected with that many problems of present transit system will be overcome. ETC (Electronic toll collection) and VICS (Vehicle Information and Communication Systems) are already used practically in Japan, and the research is advanced in order to carry out the more advanced communication. ITS is a fusion technology between the vehicles and communication, and provides drivers and passengers with comfortable and safety travelling environment. In ITS, Inter-Vehicle Communications (IVC, communications among vehicles, not depending on infrastructure of road side) and Road-Vehicle Communications (RVC) are expected to play an important role for assisting safe driving, and supporting automatic driving such as Automated Highway Systems (AHS). The quality of such system is a matter of life or death for many users of transportation systems. Therefore, real-time and robust communication must be secured for ITS.

CRL Yokosuka (Yokosuka Radio Communications Research Center, Communications Research Laboratory, MPT Japan) has intensively set up millimeter-wave test facilities in order to accelerate research activities on the ITS wireless communications.

In this chapter, we first introduce the millimeter-wave test facilities for the ITS Inter-Vehicle Communication. For the IVC experiments, we have prepared two vehicles on which experimental apparatus for the evaluations of propagation characteristics and transmission characteristics in the millimeter-wave frequency band of 60 GHz have been mounted. Using these apparatus, we have executed experiments and have obtained useful experimental results on a public road in the YRP (Yokosuka Research Park). Some results are shown in this chapter.

Then, we introduce the road-vehicle communications test facilities based on the Radio on Fiber (ROF) transmission system and micro-cell network system along a road in the YRP. In these facilities, millimeter-wave frequency bands of 36 − 37 GHz as the experimental band are used. A control station is located on the 3rd floor of a research building and 12 antenna poles for the roadside base stations are put up in the equal interval of 20 meters along an about 200 meters straight line road. Optical fiber cables are installed in the state of a star connection between the Control Station (CS) and each roadside Local Base Station (LBS). Propagation characteristics between the roadside antenna and a vehicle are presented and overall transmission system including optical fiber cable section and air section are also mentioned in this chapter.

By using Radio on Fiber transmission system like this, it is possible to transmit multiple services in one fiber. At present, a service offered to usual mobile radio communication such as TV, broadcasting of radios, etc., VICS, portable telephone is raised, and in addition, new systems such as the electronic toll-collection system will be also introduced in future. However, many mobile terminal must be purchased, since the user receives these services at present. Software radio will be able to solve these problems. It is possible that the user receives mobile communication and broadcast service of the multiple by using this system, by pass one terminal.

In this chapter, new configuration method of multimode software radio system by parameter controlled and telecommunication component block embedded digital signal processing hardware (DSPH) is proposed for the future flexible multimedia communications. In this method, in advance, basic telecommunication component blocks are implemented in the DSPH like DSP and FPGA. And, external parameters, which are simple but important information, change the specification of each block. This proposed method has the following features: i) People need to have only one mobile handset and select communication services as they like. ii) The volume of download software is reduced drastically in comparison with conventional full-download-type software radio system. iii) Since important component blocks have already been implemented into the DSPH except for some external parameters in advance, the know-how related to the implementation of DSPH never leak out. In this chapter, we evaluate the effectiveness of the proposed configuration method by using computer simulation and developed experimental prototype, and comparing with full-download-type software radio system from the viewpoint of the volume of download software. Finally, we introduce several new software radio systems by using the proposed configuration method.

2 INTER-VEHICLE COMMUNICATION [1]

2.1 EXPERIMENTAL FACILITY FOR INTER-VEHICLE COMMUNICATION

In millimeter-wave propagation between vehicles, propagation condition is affected by various factors due to environmental change as traveling of vehicles. To investigate the behavior of propagation characteristics comprehensively, we prepared an experimental facility for IVC systems using millimeter wave.

Figure 8.1 shows the block diagram of our experimental system. There are two vehicles for the IVC measurement. The precedent car has one RF section (A), and the following car equipped two RF sections (B and C) for space diversity. Each RF section has the transmitter and receiver. The propagation experiments are executed by use of the RF section (A) as the transmitter and these B and C as the receivers. The two-way data transmission is also available for the demonstration of general data transmissions such as 10 Mbps Ethernet.

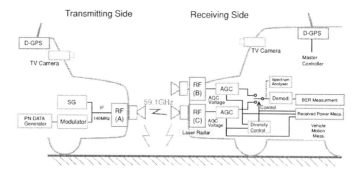

Figure 8.1   Block diagram of experimental systems.

Frequency division duplex is used for two data links. The center frequencies of RF are 59.1 GHz (for A) and 59.6 GHz (for B, C).

In the transmitting side, signal generator makes the carrier frequency of 140 MHz. This signal generator also makes various modulation signals as ASK, xFSK, xPSK, and xQAM for modulation analysis. In data transmission experiment, PN code at 1, 5 or 10 Mbps is made by the data generator, and IF carrier at 140 MHz is modulated by Manchester-DFSK in the modulator. The IF signal is upconverted to RF signal at 59.1 GHz by the RF section using MMIC devices. The RF section is in a waterproofing radome with the constant temperature control. This radome is installed at a constant height in the rear of the vehicle.

In the receiving side, two RF sections are located at the constant heights in front of the vehicle. RF signals are downconverted to IF frequency at 140 MHz in RF section. The gain of these two IF signals are controlled by the AGC section. The two AGC voltage signals that correspond to the received power, are stored at the DAT storage at a sampling rate of 150 kHz or less. One of these two IF signals are selected according to the diversity section control. The selected IF signal is demodulated, and bit error rate is measured. This IF signal is also input to real-time spectrum analyzer for the modulation analysis.

Diversity switching is triggered by comparison between each instantaneous AGC voltage with adjustable threshold value of each AGC voltage, threshold value of difference of AGC voltages, and delay timing. If the received power is less than threshold level and difference of received power is more than threshold level, the switching is executed after the constant delay. Switching method is "switch-and-stay". This diversity switching is not considered the synchronization for the bit timing. The diversity-control signal is also stored by the DAT. In the IVC using the millimeter wave, conditions of the propagation channel should be affected by the change of environmental conditions

Figure 8.2   Scenery of test course.

such as buildings or fences and a vibration of vehicles. Thus CCD camera is equipped at the front of the vehicle and digital video system which is synchronized with the other measurement systems, records visual information for various environment conditions around the vehicles. An optical gyroscope is also equipped for the measurement of instantaneous motion of each vehicle separated into three axes of gyration. The laser radar with the resolution of 2.5 cm is equipped at the front of following vehicle. This radar measures the instantaneous distance between the vehicles. In the measurement, the data of environmental conditions in both sides of the vehicles and the propagation data are synchronized by the D-GPS signal with each other. The offline data-playback system is equipped in a room. This system can play obtained data visually and synchronously, and it analyzes the propagation parameter such as distribution of cumulative probability of the received power.

2.2   EXPERIMENTS

1 Mbps wireless digital data transmission with a carrier frequency of 59.1 GHz was examined between a transmitter (Tx.A) on a fixed precedent car and two receivers (Rx.B, C) on a following car. Figure 8.2 also shows the experimental scenery. The test course is straight two-lane pavement and almost 200 m long. There are one building and several prefabricated houses, and several banks around the course. There were few objects that cause reflection, and there was no obstacle between the Tx and Rx. The precedent car was parked

Table 8.1   Experimental setup.

Center frequency	59.1 GHz
Transmitted power	-4 dBm
Data rate	1 Mbps
Modulation	DFSK (manchester code)
Detection	Differential
Antenna	Standard Horn
Antenna gain	24 dBi
Polarization	Vertical or Horizontal
Diversity threshold (Level)	-70 dBm
Diversity threshold (Diff.)	10 dB
Diversity timing delay	10 μs

at the edge of the road, and the following car moved at the constant speed of 2.5 m/s from the another edge of the road to the precedent car.

Table 8.1 shows the experimental setup for the measurement. The transmitted power is -4 dBm. Each antenna at Tx and Rxs is a standard horn antenna with the gain of 24 dBi, and these were placed at the height of 46 cm (Tx.A), 85 cm (Rx.B), and 38 cm (Rx.C) respectively. The bit error rates (BERs) were measured each one-second and the received powers were also measured simultaneously at the rate of 18750 points per second. Diversity threshold of the absolute level is set at -70 dBm, that of difference level is set at 10 dB, and timing delay is set at 10 micro seconds.

2.3   TWO-RAY MODEL FOR MILLIMETER WAVE PROPAGATION

The two-ray propagation model with a direct wave and a reflected wave from the pavement was applied for estimation of propagation characteristics of millimeter wave. Figure 8.3 is a schematic view of the two-ray propagation model. In this model, the received power P_r is expressed approximately as

$$P_r = \frac{P_t G_t G_r}{L(d)} \left(\frac{\lambda}{4\pi d}\right)^2 \sin\left(\frac{2\pi h_t h_r}{\lambda d}\right) \tag{8.1}$$

where P_t is the transmitted power, G_t and G_r are the antenna gains at the transmitter and the receiver, $L(d)$ is the absorption factor by oxygen, λ is the wave length, d is the distance between the antennas, h_t and h_r are heights of the transmitter and the receiver, respectively. In this model, the reflection coefficient of the pavement is assumed as -1 and the directivity of antenna is ignored. Absorption of oxygen is assumed as 16 dB/km.

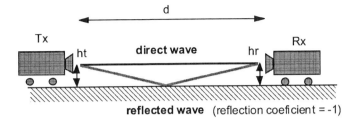

Figure 8.3   Two-ray propagation model.

2.4   RESULTS

Fig. 8.4 shows the measurement results of relationship between the received power and BER, and horizontal distance between the vehicles at each Rx position for (a) Rxh = 85 cm, and (b) Rxh = 38 cm. The estimated received power using two-ray propagation model is also indicated by dashed line in Fig. 8.4. Bit error rates are also shown in Fig. 8.4 as circular markers, where the 10^{-10} shows error free. The results of measured receiving power give fairly good agreement with those obtained by the two-ray propagation model. In this graph, it is found that the bit error rates are degraded when the received power is not sufficient.

Figure 8.5 shows the measurement result when the vertical space-diversity is applied. The received power and BER are not so much degraded as those when the vertical space-diversity is not applied. This result shows that the vertical space-diversity is effective in improving data transmission performance for IVC system using millimeter-wave experimentally.

Figure 8.6 shows the measurement result of cumulative distribution of BER travelling on the expressway. Although the shadowing by other vehicles was occurred many times, the error-free transmission was realized for the period of 81% of the travel time on the expressway.

Figure 8.7 shows the measurement results of the relationship between the received power and horizontal distance between the vehicles on the expressway. The characteristics of received power are different from those from two-ray model. This will be caused by the fluctuation of the vehicles.

3   RADIO ON FIBER ROAD-VEHICLE COMMUNICATION [2;3]

3.1   CONFIGURATION OF THE PROPOSED SYSTEM

Nowadays the number of communications equipment of the car, especially antenna, has been increased, because many services such as Vehicle Informa-

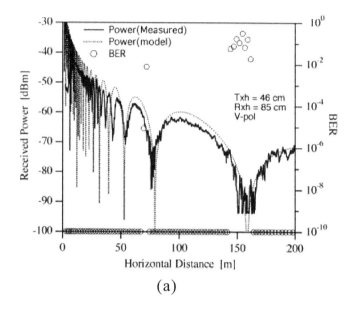

(a)

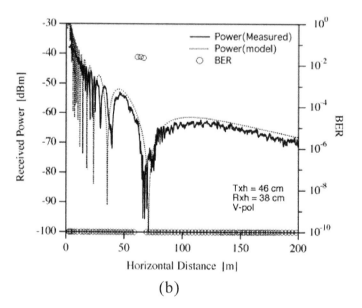

(b)

Figure 8.4   Measurement results of relationship between received power / BER and horizontal distance, (a) Rxh = 85 cm, (b) Rxh = 38 cm.

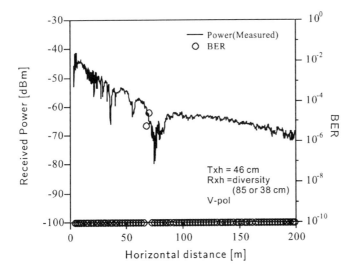

Figure 8.5 Measurement result when vertical space-diversity is applied.

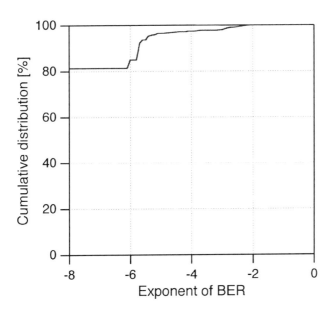

Figure 8.6 Results of cumulative distribution of BER traveling on the expressway.

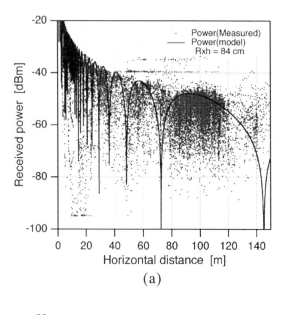

(a)

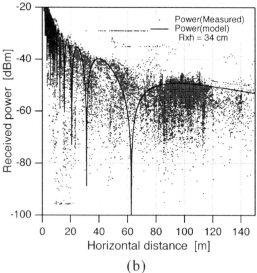

(b)

Figure 8.7 Measurement results of relationship between received power and horizontal distance traveling on the expressway, (a) Rxh = 84 cm, (b) Rxh = 34 cm.

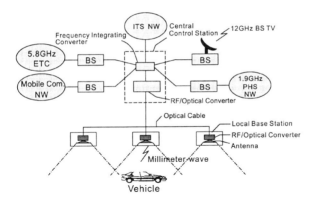

Figure 8.8   Concept of the ITS multiple service network based on CFB-ROF system.

tion & Communication System (VICS), TV and mobile communications are available on different frequency bands. As a result, the car looks like a "hedgehog". However, by using the common frequency band, the number of air interface between the car and the wireless service network is drastically decreased. This is an important factor from the view points of not only the car design but also the efficient frequency use.

Figure 8.8 illustrates the concept of the ITS multiple service network based on the Common Frequency Band Radio On Fiber (CFB-ROF) transmission. In this technique, first of all, we convert the radio frequencies of various wireless services into the common frequency band. The users of the ITS can use this common specified frequency band for the ITS multiple service communications.

For the down-link of this system, the combined electrical radio signal, which is converted to the common frequency band, drives EAM and the modulated optical signal is delivered to the Local Base Station (LBS). Then, by using Photo Detector (PD), the optical signal is converted to the radio signal and is transmitted to the vehicle from the roadside antenna.

The vehicle has only to have the antenna which matches with the common frequency band and receives the radio signal from the LBS. In the vehicle, the radio signal is converted and divided into the original band of each service. Finally, the signal is carried to each terminal on the original band by the distributor.

The distributor may be equipped with several connectors for distribution to each terminal. We can connect the distributor and each off-the-shelf terminal with a cable. If we use a multi-mode terminal, it is not necessary to distribute the received signals to each terminal. Multi-mode terminal is expected to be

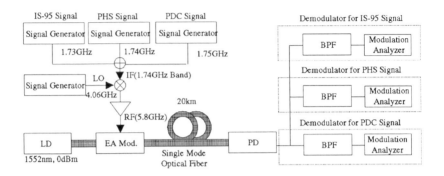

Figure 8.9   Configuration of the experimental setup.

realized by adopting the software radio technology. Furthermore, multi-mode terminal will make a contribution for the efficient space use in a vehicle. For the up-link, the procedure is the reverse of the down-link.

3.2   EXPERIMENTS

Figure 8.9 shows the experimental setup for the optical transmission of three kinds of mobile communication services, IS-95, PHS and PDC in Japan. In this experiment, 5.8 GHz band is used as the common frequency band and the interval of each carrier frequency is set at 10 MHz. The wavelength of laser diode is 1552 nm and its output power is 0 dBm. The insertion loss of electroabsorption external modulator is about 9 dB. We use four kinds of optical fiber length as almost 0 m, 5 km, 10 km and 20 km. At the receiving side, each channel of three services is filtered out after detection by the PD which has a frequency response up to about 60 GHz. After demodulation of each channel signal, the transmission quality were measured by modulation analyzer.

Figure 8.10 shows the frequency response of the ROF link of the experimental setup. Due to the chromatic dispersion of the single mode fiber, the received power decreases at every constant frequency interval depending on the fiber length. In the case of fiber length of 10 km, the first power decreasing frequency is about 15 GHz.

Figure 8.11 shows a measured spectrum of three different kinds of mobile communication channels, IS-95, PHS and PDC. As shown in Figure 8.11, high dynamic ranges were obtained. Table 8.2 shows the results of the measurements of transmission qualities of these channels. The error vector magnitude (EVM) for PHS and PDC, and the ρ for IS-95, which are standard evalua-

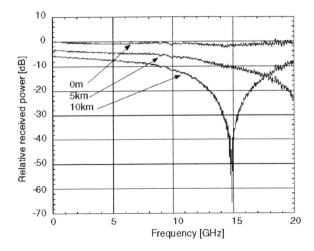

Figure 8.10 Frequency response of the ROF link of the experimental setup.

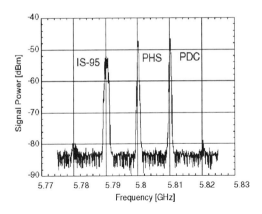

Figure 8.11 Measured spectrum of triple service radio transmission by CFB-ROF.

tion parameters, normally must be less than 12.5% and 0.99 respectively. The measured values were sufficiently good relative to these normal values.

3.3 DEVELOPMENT OF PROTOTYPE SYSTEM

We have successfully developed a prototype system for dual wireless service, e.g. ETC and PHS, utilizing CFB-ROF technique in the frequency band of 5.8 GHz region. Figure 8.12 illustrates the configuration of the developed

Table 8.2   Measurement results of modulation qualities.

Fiber length	IS-95 (ρ)	PHS (EVM:%)	PDC (EVM:%)
5 km	0.99764	2.30	2.28
20 km	0.99750	3.07	2.29

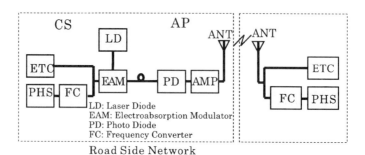

Road Side Network

Figure 8.12   Configuration of the developed prototype system.

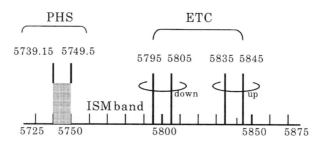

Figure 8.13   Frequency allocation of PHS and ETC in the prototype system.

system. This system consists of a roadside network and a mobile terminal in the vehicle. This prototype can support the two-way communications.

The original frequency band of PHS is in 1.9 GHz region and it is converted into 5.8 GHz region. Figure 8.13 illustrates the frequency allocation for PHS and ETC in this system. By using this prototype with PHS handsets and a set of ETC terminal and server, we can get an announcement of the toll gate charge

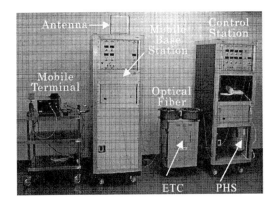

Figure 8.14   Overview of the prototype for ITS and dual service radio transmission system in 5.8 GHz band.

Table 8.3   Specificatin of the developed system.

Modulation	PHS	$\pi/4$ DQPSK (384 kbps)/TDMA-TDD
	ETC	ASK (1.024 Mbps)/slotted ALOHA
Frequency	PHS	5739.15-5749.95 MHz
	ETC	5795, 5805 (down), 5835, 5845 (up) MHz
Antenna Gain	Road	18 dBi
	Vehicle	5 dBi
Output RF Power	Road	3 dBm
	Vehicle	10 dBm
Optical fibe Length	1 km	

and can make a phone simultaneously. Figure 8.14 shows an overview of the prototype system for PHS and ETC and Table 8.3 shows its specifications.

3.4   EXPERIMENTAL FACILITIES FOR RVC IN 37GHZ BAND

Figure 8.15 shows the configuration of the experimental facilities for the ROF transmission system of three kinds of mobile services such as Electric Toll Collection (ETC), Personal Handy Phone (PHS) and TV broadcasting in Japan. In these experimental facilities, 37 GHz band is used as the common frequency band and the frequency for each service is allocated as shown in

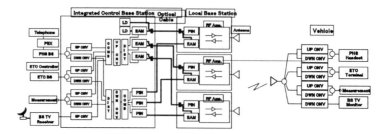

Figure 8.15 Configuration of the experimental facilities for the multiple service CFB-ROF transmission system.

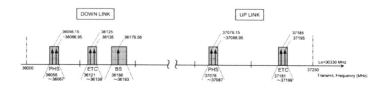

Figure 8.16 Frequency allocation of multiple service ROF transmission system in 37 GHz band.

Figure 8.16. The wavelength and the output power of the laser diode for this scheme is in 1.5 μm region and about 0 dBm, respectively. The length of optical fiber cable section between Control Station in the research building and the roadside LBS is about 700 meters. The frequency bands for the down link and up link are 36.00 − 36.50 GHz and 36.75 − 37.25 GHz, respectively. The frequency bands of the RF amplifiers and antennas for the LBS and the vehicle match with these frequency bands.

We have prepared two kinds of roadside antenna. One is the horn antenna and the other is the patch antenna with 20 element antennas and both of them have the cosec-squared beam pattern on the vertical plane. The interval between roadside antennas is 20 meters. The antenna, the frequency converter and the mobile terminals are mounted in the vehicle. The original frequency bands of PHS and ETC are in 1.9 GHz and 5.8 GHz region, respectively. Therefore, the received RF signals are divided and delivered into the each mobile terminal after frequency down conversion in the vehicle. Figure 8.17 is a photo of the Control Station and the roadside antennas are shown in Fig.8.2.

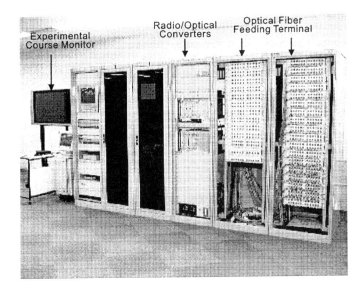

Figure 8.17 Facilities for ROF Control Station.

3.5 ESTIMATION OF RECEIVED POWER FOR ROF ROAD-VEHICLE COMMUNICATION

Next, we estimate the received power at the Mobile Station (MS). In this system, three LBSs connected to one CS transmit the same frequency radio wave to the vehicle. It is predicted that the strong interference can be observed at the boundary area between two cells covered by different LBSs.

In this simulation, each LBS is on the 5 m-height of the pole installed along the road and the poles at the roadside are lined 20 meters interval. The height of vehicle antennas is 2.1 meters. So the height deference between LBS and MS antennas is 2.9 meters. In this estimation, the transmitted power is 10 dBm and the frequency is 36.06155 GHz. The transmitting antenna has cosec-squared beam pattern on the vertical plane. This antenna enables us to get almost the same received power in the coverage area. The receiving antenna on the vehicle has pencil beam pattern with 3 dBi gain. We did not consider the reflection from road or other objects.

Figure 8.18 (a) shows the contour map of calculated received power of 5.0 m x 40.0 m area on the road. Antenna poles stand in the 20 meters interval along the roadside. Figure 8.18 (b) shows the received power at 2.1 m height from road surface and on the center of lane, i.e., 2.5 m from edge of road. The variation of the received power as a function of position is caused by the inter-ference of the radio waves from several LBSs. The interference between LBSs

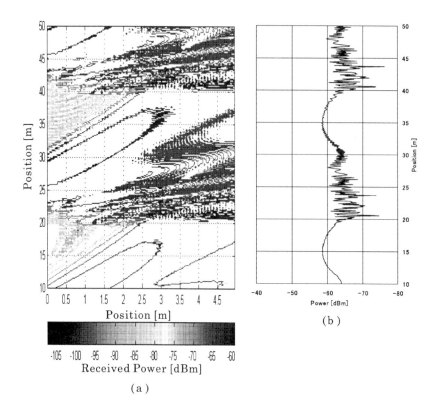

Figure 8.18   (a) Contour map of calculated received power, and (b) received power variation on the center of the lane.

causes very complicated fluctuations of received power. This result shows that we need to develop some new technologies, for example, some kind of diversity with very high-speed signal-selection.

3.6   SUMMARY

In this section, we have newly proposed an integration method of wireless multi services in ITS, which is based on Common Frequency Band Radio On Fiber (CFB-ROF) technique. Moreover, we have confirmed feasibility of our proposed system. The CFB-ROF will become a key technique for the mobile multimedia communications in ITS. As a further study, we here open a new concept for ITS services, which we have named MLS (multimedia lane and station). Figure 8.19 shows the concept of MLS. MLS consists of multime-

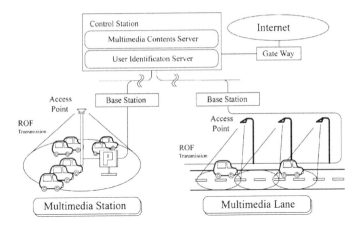

Figure 8.19   Concept of Multimedia lane and Station (MLS) proposed by Communications Research laboratory.

dia lanes and stations which provide multimedia communication services to cars moving on a road and to cars stopping at a place such as a parking lot, respectively.

4   SOFTWARE RADIO

4.1   CONCEPT OF THE PROPOSED SOFTWARE RADIO SYSTEM

4.1.1   Configuration of the conventional software radio.   In most case of software radio system,s several software programs, which describe all telecommunication components in DSPS (digital signal processing software) language, are downloaded to the DSPH (digital signal processing hardware) and it configures the components on the DSPH. And by changing the software, we can realize our required system. The software radio system can be called as full-download-type software radio system.

Figure 8.20 indicates the configuration of full-download-type software radio system. Mostly, the configuration of the full-download-type software radio system is categorized into three units: RF (Radio Frequency) unit, IF (Intermediate Frequency) unit and baseband unit. These are called as RFU, IFU and BBU, respectively in this paper. RFU handles antenna block, up- and down-converter blocks. IFU consists of quadrature modulator and quadrature demodulator blocks, A/D (Analogue to Digital) and D/A (Digital to Analogue) converter blocks. On the other hand, BBU consists of several baseband DSPH like DSP or FPGA. These DSPH can change the specification by changing

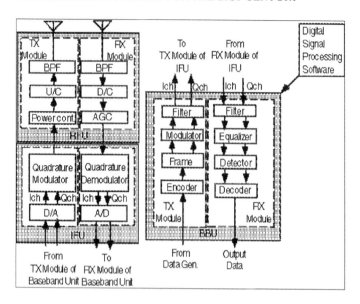

Figure 8.20   The configuration of full-download-type software radio system.

software, which can configure all telecommunication components related to transmitter and receiver. These three units also have two parts: TX module and RX module. TX module has relation to the transmitter. On the other hands, RX module is related to the receiver.

When we would like to realize a telecommunication system by full-download-type software radio system, first of all, all DSPS, in order to configure a required telecommunications system, are downloaded to BBU before starting communication. As for the software, frame block, encoder block, mapping and modulation block, filter block and so on are described in DSPS language and downloaded to the BBU of TX module. Moreover, filter block, equalizer block, detector and decoder block are also described in DSPS language and downloaded to the BBU of RX module. After finishing the download of software, the configuration check program is executed. Finally, BBU configures the required baseband modulation and demodulation circuit. Then, transmission data are fed into BBU of TX module.

In the BBU of TX module, the input transmission data are framed modulated and converted to two signals: In-phase channel (Ich) and Quadrature-phase channel (Qch) signals by several DSP blocks mentioned above. Then, the digitally modulated data are fed into IFU of TX module.

In the IFU of TX module, the digitally modulated Ich and Qch signals are converted from digital data to analogue data by a D/A converter block. Then, the converted analogue Ich and Qch signals are quadrature modulated on the IF band and send to RFU of TX module.

In the RFU of TX module, the quadrature modulated signal on the IF band is up-converted to RF band through power control part and finally transmits to the air.

On the other hand, when we receive the RF signal, first of all, the received signal is fed to the RFU of the RX module. In the RFU of RX module, the received RF signal is bandpass-filtered to eliminate spurious signal and down-converted to the IF band. Then, Automatic Gain Control (AGC) block controls the power of the down-converted signal in order to keep constant level to A/D converter in IFU of RX module. Afterwards, the power-controlled signal is fed to the IFU of RX module.

In the IFU of the RX module, the received signal from RFU is split into two signals: Ich and Qch signals through a quadrature demodulator block. Then, by using A/D converter block, these split signals are over-sampled and transferred to BBU of the RX module.

In the BBU of RX module, all telecommunication component blocks have been implemented by the DSPH before starting communication by changing the download software for DSPH, and the configuration has been checked by test program. Therefore, the Ich and Qch over-sampled signals are filtered, equalized by customize method, detected and decoded by using filter, equalizer, and decoder blocks in the BBU of RX module. Finally we can recover the transmission data.

In the conventional full-download-type software radio, we can change the system configuration in accordance with our request easily. However, the following problems are involved.

1. The volume of software downloaded into the DSPH increases, as the contents of the required telecommunication component blocks become more complicated. As a result, the download time is lengthened. In addition, redundancy code for coding must be added for the download software, when the concealment of the download program or the robustness for all jamming signals or fading is considered. In this case, the download time is lengthened more and more.

2. The period for configuration check of the DSPH also increases, as the contents of the required telecommunication component blocks become more complicated. The problem also comes out in the stability of the operating characteristic of the DSPH, when we can't have sufficient period for the configuration check.

3. In the download software, there are often several component blocks which are related to the know-how of the manufacturer, e.g. the optimization method or calculation algorithm for some special blocks, etc..

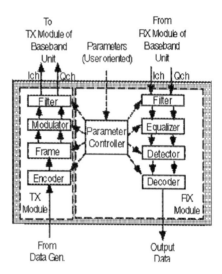

Figure 8.21   The BBU configuration of parameter-controlled-type software radio system.

The know-how may leak out by the download of software. And, there is the possibility altered from the other people.

In this section, we propose a new configuration method of software radio system to overcome these problems. In next subsection, we explain the configuration in detail.

4.1.2   Configuration of the proposed software radio system [4].   The proposed system is only related to the BBU of both TX and RX modules. For the other units, RFU and IFU, we may prepare these units for each system individually or we may integrate RFU and IFU for all systems. The BBU configuration of the proposed system is shown in Fig. 8.21. In the BBU, basic telecommunication component blocks like encoder, frame, modulator, filter blocks of transmitter and equalizer, detector and decoder blocks of receiver have already been programmed and implemented in the DSPH in advance. And the functions of the telecommunication blocks are not fixed but programmable and changeable easily by downloading external parameters. Namely, if we would like to change the configuration of digital filter of the transmitter, we send only coefficient information of the required digital filter to the filter block of BBU. By using these coefficient data, a new filter block is configured. Then, the proposed BBU unit becomes one of general-purpose transmitter and receiver by external parameter. The software radio system by a new configuration method of BBU is called as parameter-controlled-type software radio system.

In BBU of TX module, the frame format, the shape of transmitter filter, and the modulation scheme of the required system are determined on the basis of external parameters downloaded from the outside of the DSPH. By using the configured telecommunication blocks with parameters, the transmission data is modulated.

In the BBU of the RX module, the time synchronization method, the shape of receiver filter, the equalization scheme, and demodulation scheme are determined on the basis of external parameters downloaded from the outside of DSPH. By using the configured telecommunication blocks, the received signal is demodulated.

The proposed system is realized by downloading external parameters, which is small and important information for the realization of a required telecommunication system. Therefore the important blocks have been configured in the DSPH in advance. As a result, it is expected that we can obtain stable performance and reduce the period for configuration check in comparison with the full-download-type software radio system. Moreover, the information related to the know-how of manufacturer such as the optimization methods for the specified telecommunication components never leak out, because we only download a general and small volume software. Moreover, since the volume of download software to the DSPH is not quite small, we can utilize several strong coding methods to the download software. Consequently, the proposed software radio communication system can keep high concealment.

4.2   PERFORMANCE EVALUATION OF THE PROPOSED SOFTWARE RADIO BY DEVELOPED PROTOTYPE

4.2.1   Configuration of the developed prototype.   In order to show the effectiveness of the proposed software radio system, an experimental prototype was developed and its transmission performance was evaluated. The following are shown in Figs 8.22 and 8.23: Appearance and system configuration of the experimental prototype. Moreover, the system parameters are summarized in Table 8.4. The experimental prototype can make use of three real telecommunication systems: PHS and GPS and ETC systems as Service mode. Moreover, as the User mode, it is possible that the user freely conducts several modulation schemes, GMSK, $\pi/4$ QPSK, BPSK and QPSK. As for PHS and GPS, we integrate the antennas of two systems into one antenna because the frequency utilized in GPS (1.5 GHz band) is close to PHS band (1.9 GHz band). And, the external parameters, which need to change the system, are supplied from a notebook type computer connected with experimental prototype by the

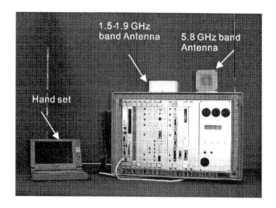

Figure 8.22   Appearance of the experimental prototype.

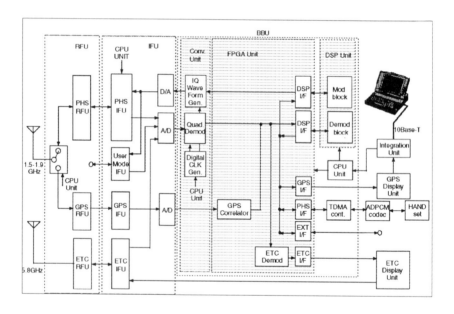

Figure 8.23   System configuration of the experimental prototype.

10Base-T Ethernet cable. In the notebook computer, management software of the experimental prototype has been installed. And we can select several services and modulation schemes of the User mode by using menu window shown in Fig. 8.24 (Service mode) and 8.25, (User mode) respectively.

Table 8.4 System parameters of the developed experimental prototype.

Services	(Service mode) PHS, ETC, GPS (User mode) BPSK, QPSK, π/4 QPSK, GMSK	Frequency	PHS — 1.9 GHz, ETC — 5.8 GHz, GPS — 1.5 GHz, User Mode — IF band
Modulation	PHS — π/4 DQPSK ETC — ASK GPS — BPSK + Spread spectrum User mode BPSK, QPSK, π/4 QPSK, GMSK	Data rate	PHS — 384 kbps(carrier bit rate) ETC — 1024 kbps: 2048 kbaud (data rate) GPS — 50 bps (data rate): 1023 Mcps(chip rate) User mode GMSK — 270.833 kbps (carrier bit rate)

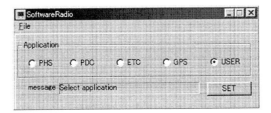

Figure 8.24 Menu window of management software on PC (Service mode).

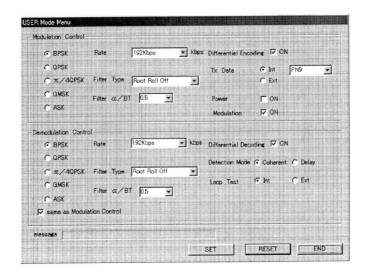

Figure 8.25 Menu window of management software on PC (User mode).

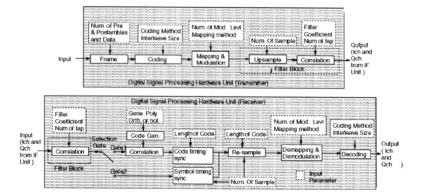

Figure 8.26   The configuration of BBU in the computer simulation.

When user would like to realize PHS system by using the configuration shown in Fig. 8.23, the external parameters which is necessary for constructing the PHS system, is stored at the CPU unit through 10Base-T Ethernet cable from a notebook type computer. The CPU unit gives the parameters to the BBU, IFU, and RFU. Then, this experimental prototype becomes the PHS mode. Afterwards, the transmission data are encoded in the handset and a speech coding (ADPCM) units. Then, the encoded data is fed to a modulation block of the DSP unit through a TDMA control unit and a PHS interface unit with DSP unit. The modulation block has the similar composition to Fig. 8.26, where all key functions are programmed in advance, and it realizes the PHS modulation scheme by changing parameters from CPU unit. Then, modulated data are filtered at the conversion unit and fed to IF unit. Then, the digitally modulated signal is converted from digital modulation signal to analogue modulation signal at IF unit, and finally it is transmitted to the air from the antenna through RF unit.

On the other hand, in the receiver, the received PHS signals firstly passed through RF unit, IF unit, and conversion unit of BBU, and are converted to quadrature demodulated digital over-sampled signals of Ich and Qch. These signal are fed to the demodulation block of DSP unit on BBU. Then, the demodulation block also has the similar composition to Fig. 8.26, where all key functions are programmed in advance, and it realizes the PHS demodulation scheme by changing parameters from CPU unit. Then, demodulated data are transferred to a TDMA control unit and a speech decoding (ADPCM) unit via a PHS interface unit with DSP unit, converted to the voice signal, and finally we can communicate the other person by using the handset unit.

Table 8.5 The volume of software for FPGA based units in developed experimental prototype.

Unit	Conv. Unit (gate)		FPGA Unit (gate)							FPGA Total gate
Device (No)	EPM 7064 (1)	EPM 7064 (1)	EPM 7064 (1)	EPF 10K20 (1)	EPM 7128 (1)	EPM 7128 (2)	EPM 9400 (2)	EPM 7192 (1)	EPF 10K250 (1)(2)(3)	
Max Gate	1250	8000	1250	20000	2500	2500	8000	3750	250000 × 3	
User Mode	1013*	2176*			725*	826*	2520*	776*		8036
PHS	1013*	2176*	113	7200	725*	413*	2520*	388*		14548
GPS							560	86	547500	54816
ETC		1088								1088
Total	1013	3264	113	7200	725	826	3080	862	547500	564583
Purpose	TX Filter (1013, shared)	RX Filter (2176 shared) ETC Demod + ETC I/F (1088)	PHS I/F (7313)		DSP IF TX (725 shared)	DSP IF TX (413 shared) EXT I/F (413)	DSP IF RX (2520 shared) GPS I/F (560)	DSP IF RX (388 shared) EXT I/F (388) GPS I/F (86)	GPS Correlator	

* means the software is shared for the realization of some system.

The similar technique can be utilized for the GPS and ETC systems. However, in the case of ETC and GPS, digital signal processing speed of some DSP blocks is in excess of several 10Mbps, e.g. correlation part of GPS system and decoding unit of ETC. Therefore, these high speed digital signal processing part are programmed in not DSP but FPGA. And by communicating between FPGA unit and DSP unit, the BBU is configured.

Moreover, in the User mode, we can change several modulation schemes as shown in Table 8.4. However there is no RF unit. Therefore, in this paper, by connecting input port and output port of User mode IFU block on IFU, we check loop-back performance from input port to output port of EXT I/F block on BBU and obtain the transmission performance: e.g. BER.

4.2.2   Comparison between full-download-type and proposed software radio.   In order to evaluate the proposed software radio system, the following comparisons are carried out between full-download-type software radio system and the parameter-controlled-type software radio system: Volume of download program and length of download time. In the prototype, as for the DSPH, FPGA by Altera Corp. is utilized in the conversion and FPGA units and DSP is used in the DSP unit. As the first results, the volume of software which can be utilized in the conversion, FPGA and DSP units are shown in Table 8.5 and 8.6. For FPGA based units, we show the software volume as the required number of gate (gate). On the other hands, for DSP based unit, we show the software volume as the volume of programmed code for DSP (byte).

Table 8.5 shows the required number of gate for FPGA which utilized in each function block and each hardware devices. In Table 8.5, * means that

Table 8.6   The volume of software for each system in developed experimental prototype.

Device	Conv. & FPGA Unit (gate)	DSP Unit (byte)			Parameter (byte)
		TX	RX	Total	
User Mode	8036	13357*	40509*	53866	1660
PHS	14548	9586*	13340*	22926	1616
GPS	548146	Not used	13340*	13340	4
ETC	1088	Not used	Not used	0	570
Total	564583	13857	40509	53866	

* means the software is shared for the realization of some systems.

the data are shared by several systems. For example, TX filter block is shared with User mode and PHS mode, and the size of FPGA is 1013 gates and programmed in EPM7064(1). From Table 8.5, we can easily understand that we need 8036, 14548, 548146, 1088 gates for the individual realization of User mode, PHS, GPS and ETC systems, respectively by full-download-type software radio system. And in the prototype, 564583 gates is pre-programmed in FPGA based units for the realization of the proposed software radio system. By the same procedure, we can investigate the volume of software for DSP unit in both case: one is needed for the proposed software radio system, and the other is required for the individual realization of User mode, PHS, GPS and ETC systems. These data are arranged in Table 8.6. In Table 8.6, we show the volume of parameters in the case of the proposed software radio system.

If we assume that we only download the software for DSP unit, the download software in the proposed software radio system becomes about 1/15(for PHS)-1/30(for User mode) in comparison with the case that we download all DSP software to DSP unit for the realization of each system. Moreover, the followings are clarified from Table 8.6. If we realize PHS terminal by full-download-type software radio system, we need 14548 gates for FPGA. However, in the proposed software radio system, we must pre-program 564583 gates in FPGA. On the other hand, for the volume of DSP software, we need 22926 byte in order to realize only PHS system by full-download-type software radio system. However, in the proposed software radio system, we must pre-program 53866 byte. Namely, we need some redundant programs for some systems. However, by the redundancy, the proposed system reduce the period of software download and obtain stable performance in the case of reconfiguration of systems.

In addition, we compare two software radio systems from the other viewpoint. That is the required scale of DSPH for the realization of both software

Table 8.7   The average configuration time in User mode by parameter-controlled-type software radio system.

BPSK ->QPSK	37.5 ms	BPSK ->GMSK	30.5 ms	QPSK -> π/4 QPSK	30 ms
QPSK ->BPSK	32.5 ms	GMSK ->BPSK	32.5 ms	π/4 QPSK ->QPSK	35 ms
BPSK -> π/4 QPSK	32.5 ms	QPSK ->GMSK	30 ms	π/4 QPSK ->GMSK	30 ms
π/4 QPSK ->BPSK	35 ms	GMSK ->QPSK	40.5 ms	GMSK -> π/4 QPSK	35 ms
Full Download User mode	3410 ms				

radio systems. From Table 8.6, for the proposed software radio system, we need 564583 gates for FPGA. On the other hand, for full-download-type software radio, the maximum volume of gate in order to realize each system, is at least needed. That is the case of GPS and the volume is 548146 gate. Therefore, in the case of the configuration of the prototype, we need FPGA unit for the proposed system which total number of gates becomes 1.02 times in comparison with full-download-type software radio. On the other hand, as for DSP unit, in this prototype, we need only the programs for User mode, because the other systems utilize some parts in the programs. Therefore, in both software radio systems, we need to prepare DSP components which can include the volume of programs for User mode.

Moreover, the average download time, when we change the modulation scheme from one system to another system, is shown in Table 8.7 by using User mode and parameter-controlled-type software radio system. In User mode, we can realize GMSK, π/4 QPSK, BPSK or QPSK by changing parameters. In addition to the above information, the download time is also mentioned in Table 8.7 in the case of full-download-type software radio for User mode. As shown in Table 8.7, the average download time of the proposed software radio system becomes around 1/100 in comparison with full-download-type software. In addition, the average download time for all modulation schemes for the realization of User mode is independent on the modulation schemes and almost same by parameter-controlled-type software radio system. This is because the download software consists of the parameters described before, and the parameters are needed to all modulation schemes and the volume is almost same for all modulation scheme for the realization of User mode.

Finally, the BER performance of this experimental prototype in the case of user mode is shown in Fig. 8.27. In Fig. 8.27, we also include a theoretical BER value in terms of E_b/N_0 for BPSK, QPSK and π/4 QPSK. Since we adopt the coherent detection, the theoretical BER value becomes $0.5\mathrm{erfc}(\sqrt{E_b/N_0})$ [5]. As shown in Fig. 8.27, it is clear that the BER performance of the experimental prototype agrees well with the theoretical value within 1 dB.

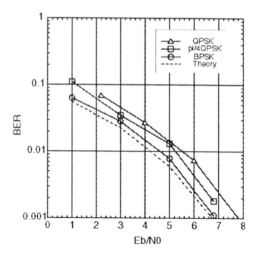

Figure 8.27   BER performance of the experimental prototype in the case of User mode.

References

[1] A. Kato, K. Sato M. Fujise, "Experiments of 156Mbps Wireless Transmission using 60GHz band on a pavement", in proc. of the First International Symposium on Wireless Personal Multimedia Communications (WPMC'98) pp.389-392, Nov. 1998.

[2] M. Fujise, H. Harada, "Multimode DSRC by Radio On Fiber", in proc. of the 1998 Communication Society Conference of IEICE, SAD-2-8, pp.32-33, Sep.1998.

[3] M. Fujise, K. Sato and H. Harada, "New Road-Vehicle Communication Systems Based on Radio on Fiber Technologies for Future Intelligent Transport Systems (ITS)", in proc. of the First International Symposium on Wireless Personal Multimedia Communications (WPMC'98), pp. 139-144, Nov. 1998.

[4] H. Harada, Y. Kamio and M. Fujise, "Multimode Software Radio System by Parameter Controlled and Telecommunication Component Black Embedded Digital Signal Processing Hardware," IEICE Trans. Commun., vol. E83-B, No. 6, June 2000.

[5] Y. Saito, *Modulation and demodulation scheme for digital radio communication system* (ISBN4-88552-135-1), p.115, IEICE 1996 (in Japanese).

Chapter 9

WIRELESS DATA COMMUNICATIONS SYSTEMS

Kaveh Pahlavan
Center for Wireless Information Network Studies, Worcester Polytechnic Institute, USA
kaveh@ece.WPI.edu

Xinrong Li
Center for Wireless Information Network Studies, Worcester Polytechnic Institute, USA
xinrong@ece.WPI.edu

Mika Ylianttila
Centre for Wireless Communications, University of Oulu, Finland
mika.ylianttila@oulu.fi

Matti Latva-aho
Centre for Wireless Communications, University of Oulu, Finland
matla@ees2.oulu.fi

Abstract   The wireless data communication industry has experienced fast development in the past few years. With the finalization of new series of IEEE 802.11 and ETSI BRAN HIPERLAN standards, new features have been integrated into the conventional wireless LAN, which was introduced as an alternative of fixed LAN. New emerging technologies, such as HomeRF and Bluetooth, are becoming new impetus for the fast expansion of the market. In this paper, we present an overview of the current status and future trends of wireless data communication systems.

Keywords: Wireless LAN, IEEE 802.11, BRAN HIPERLAN/2, HomeRF, and Bluetooth

1   INTRODUCTION

The allocation of worldwide available unlicensed radio spectrums at 2.4 GHz ISM (Industrial, Scientific and Medical) band and 5 GHz U-NII (Unlicensed National Information Infrastructure) band has prompted dramatic interests in the wireless industry to develop broadband wireless data communication systems. Starting from the wireless LAN technology, more and more technologies and applications are developed for expanding the market [1, 2].

The wireless LAN (WLAN) industry has been expanding continuously into the health care, manufacturing, finance, small business and educational markets [3]. With the finalization of new series of IEEE 802.11 and HIPERLAN standards, wireless LAN is evolving into very high-speed data transmission supporting both packet- and connection-oriented voice, Quality of Service (QoS), etc [4]. These new trends of wireless LAN technology will certainly become the impetus for the fast development of markets. However the slow deployment of wireless LANs in the past years has created opportunities for other technologies such as Bluetooth and HomeRF. A wide consensus to date is that there is no specific wireless "killer" application beyond mobile telephone services and mobile Internet access. However, with the successful emergence of new short-range radio technologie (e.g. HomeRF, Bluetooth), the wireless industry is believed to have found a right way for a healthy evolution.

In this paper we provide an overview of the current status and trends of wireless data communication systems. The paper is organized as follows. In the second section, we describe wireless data application scenarios and current status of the market. In Section 3, we present an overview of four wireless data communication standards – IEEE 802.11, HIPERLAN/2, HomeRF SWAP and Bluetooth. In Section 4, challenges and problems behind wireless data communication systems as well as future trends are presented. Finally, conclusions are given in the last section.

2   APPLICATIONS AND MARKETS OF WIRELESS DATA COMMUNICATION SYSTEMS

Wireless LAN

Exactly as the name implies, the wireless LAN is a local area network implemented without wires. This means that all the functionalities of a conventional fixed LAN are available in a WLAN including file sharing, peripheral sharing, Internet access, etc, as shown in Figure 9.1 [2]. The WLAN can be such implemented to replace or to extend the capability of the LAN by providing mobility. Compared to the fixed LAN, the main advantages and benefits of wireless networks are the mobility and cost-saving installation [3]. Most of the application scenarios of WLAN are related to these two features. Mobility enables users to

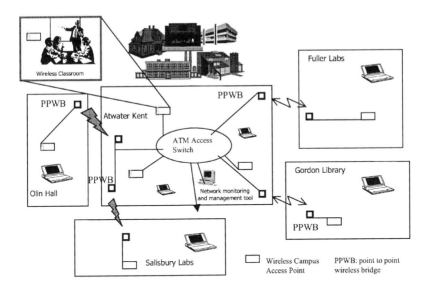

Figure 9.1   Wireless campus network at Worcester Polytechnic Institute.

roam about while being connected to backbone networks. Many jobs require workers to be mobile, such as inventory clerks, healthcare workers, police officers, and emergency-care specialists. Wireless networking provides significant cost savings in the areas where cables cannot be easily installed, such as historical buildings and residential houses. In remote sites, branch offices and other situations where on-site networking expertise might not be available or fast networking is needed, computers equipped with wireless LANs can be pre-configured and shipped ready to use. Conclusively, wireless networking is applicable to all situations where mobile computer usage is needed and/or the cable installation is not feasible.

Since the beginning of the nineties, proprietary WLAN products for ISM bands have appeared in the market. In 1997, IEEE 802.11 was standardized for 2.4 GHz band, supporting 1 Mbps and optionally 2 Mbps. In 1998, a higher speed extension to 802.11 was approved providing 11 Mbps throughput. Nowadays the 11 Mbps 802.11-based products are dominating the WLAN market. The emergence of broad market and wide deployment of WLAN (i.e. 'the year of WLAN') has been expected for so many years to date. But the development of WLAN market has been held back due to issues such as poor marketing, relatively high price and relatively low throughput (especially when compared to the wired counterparts). Most recently with the standardization

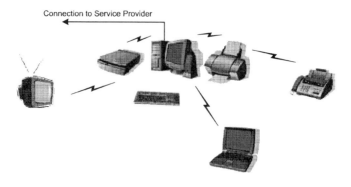

Connection to Service Provider

Figure 9.2   Applications of Wireless Home Networking (HomeRF).

of HIPERLAN/2 and IEEE 802.11a and 802.11b, a new fast development of market is being expected again.

Wireless Home Networking [6]

The wireless local area network business has been focused on offices since the industry began. But recently, home networking is seen to be a fast growing market. The personal computer has become a powerful platform for education, entertainment, information access and personal finance applications. At home, with the wide deployment of PCs at home and the Internet becoming the main way to access information, the role of the PCs will expand especially in the area of communication. More and more families have multiple PCs which gives rise to the need for home networking to share files and printers and to access Internet through only one access point (modem or cable modem). The home electronic devices are all becoming more and more intelligent with built-in computing and communication capabilities. The power of these built-in capabilities cannot be utilized while remaining isolated. All above indicates a growing need for more effective management and integration of communications between PCs and intelligent devices in homes. Continuous expanding capabilities of PC makes it a potential powerful control platform in the home and the difficulty of wiring in residential places leads to the concept of wireless home networking as shown in Figure 9.2.

The HomeRF consortium is an industry working group including major players in the PC industry (Compaq, Hewlett-Packard, IBM, Intel and Microsoft), and in the wireless telecommunication and consumer electronic industry (Ericsson, Motorola, Philips, Proxim and Symbionics). The mission of the HomeRF working group is described as "To bring about the existence of

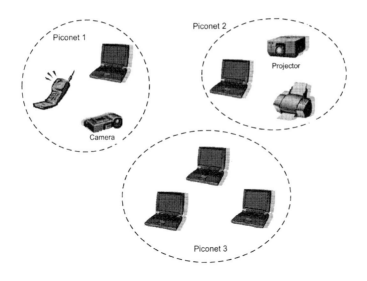

Figure 9.3 Applications and Network Architecture of Bluetooth technology.

a broad range of interoperable consumer devices by establishing open industry specifications for unlicensed RF digital communications between PCs and consumer electronic devices anywhere in and around the home". The HomeRF standard supports a broad range of new home networking applications: shared access of Internet connections from anywhere in the home, automatic routing of incoming telephone calls to one or more cordless handsets, fax machines or voice mailbox, home wireless LAN to share files, programs and printers among multiple PCs and control of home security systems, heating and air conditioning systems from anywhere around the home.

Bluetooth [8]

Bluetooth is an emerging short-range networking technology developed for 2.4 GHz ISM band. The objective of Bluetooth technology is to replace cables and infrared links used to connect disparate electronic devices with one universal short-range radio link. It also provides a mechanism to form small-scale ad hoc wireless networks of electronic devices, supporting 1 Mbps communication capability over a short-range (about 10m), as they come within the range of each other. Figure 9.3 shows some applications and the ad hoc piconet architecture of Bluetooth. Most of Bluetooth applications reflect the mobile phone industry background of the inventors (Ericsson, Nokia and etc.) of Bluetooth technology. However, as time goes by, more and more applications in various industry segments will be created for Bluetooth.

Some examples of the applications created based on Bluetooth technology are as follows [8]. Three-in-one phone: when the user is in the office, the phone functions as an intercom (no telephony charge); at home, it functions as a portable phone (fixed line charge); and when on the move, the phone functions as a mobile phone. A user can use a laptop to surf the Internet wherever the person is, and regardless if he (or she) is cordlessly connected through a mobile phone or through a wire-bound connection. In meetings and conferences, users can share information instantly with all participants without any wired connections. A user can also cordlessly run and control, for instance, a projector, or can connect his headset to his laptop or any wired connection to keep hands free for more important tasks while in the office or in the car. When laptop receives an email, the user will get an alert on mobile phone. Users can also browse all incoming emails and read those selected on the mobile phone's display. A user can cordlessly connect his camera to his mobile phone or any wire-bound connection, and also connect all peripherals to a PC or to a LAN.

These kinds of application scenarios extend the usage of any equipment or device equipped with Bluetooth radio link. According to a report from global research firm Frost & Sullivan, the market of Bluetooth is predicted to be $36.7 million in 2000 and $699.2 million by 2006 [13]. The widespread industry support for the technology (including more than 1000 companies in the Bluetooth Special Interest Group) is believed to be the main force to ensure the successful future of this new technology.

3   WIRELESS DATA COMMUNICATION STANDARDS

Currently, four standards for wireless data communications systems are available, IEEE 802.11, ETSI HIPERLAN standards, HomeRF SWAP and Bluetooth specifications. In this section, brief technical overviews of these standards are presented.

3.1   THE IEEE 802.11 STANDARDS [3]

The IEEE Standard for Wireless LAN Medium Access (MAC) and Physical Layer (PHY) Specifications, which is also known as IEEE 802.11, defines over-the-air protocols necessary to support networking in a local area. The 802.11 standard provides MAC and PHY functionality for wireless connectivity of fixed, portable, and moving stations at pedestrian and vehicular speeds with a local area. There are two possible architectures for a WLAN under the IEEE 802.11 specification as shown in Figure 9.4. The stations can communicate directly with each other in ad hoc networks. Such a configuration is also referred to as an independent configuration. There is usually no connection to the wired network. In access point (AP) based networks (or infrastructure

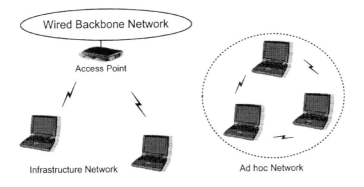

Figure 9.4   Architectures of IEEE 802.11 WLAN.

networks) the mobile terminals(MT) communicate directly with an AP that is connected to the wired network. Each AP serves a coverage area called a basic service set (BSS). Multiple BSSs form an extended service set (ESS). The inter access-point protocol (IAPP) is used for communicating between different APs in an ESS for handoff related purposes.

The IEEE 802.11 standard provides two physical layer specifications for RF, operating in 2.4 GHz ISM band, and one for infrared. For both Frequency Hopping and Direct Sequence Spread Spectrum physical layers, two different data rates are specified, 1 Mbps and optional 2 Mbps. The basic access method of 802.11 MAC is a scheme called Carrier Sense Multiple Access with Collision Avoidance (CSMA/CA). Before transmitting, a station senses the channel. When the channel is idle, the packet is transmitted right away. If the channel is busy, the stations keep sensing the channel until it is idle, then waits a uniformly distributed random backoff period before sensing the channel again. If the channel is still idle it transmits its packet, otherwise it backs off again. The backoff mechanism results in the avoidance of the collision of packets from multiple transmitters who all sense a clear channel at about the same time. All directed traffic receives a positive acknowledgement and packets are retransmitted if an acknowledgement is not received.

Shortly after the 1 Mbps and 2 Mbps standards were approved, 802.11b and 802.11a working groups started working on higher-rate extensions of the physical layer at the 2.4 GHz ISM band and 5.2 GHz U-NII band respectively. In July 1998, new rate extension, 5.5 and 11 Mbps, for 2.4 GHz ISM band is adopted for providing multi-rate operations at 1, 2, 5.5 and 11 Mbps. The draft standard of 802.11a is based on OFDM (Orthogonal Frequency Division Multiplexing) modulation scheme that was selected for its robustness against frequency selective fading and narrowband interference. The specifications of

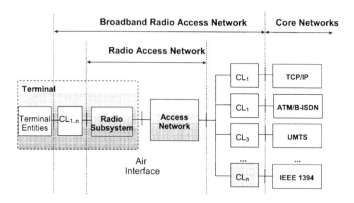

Figure 9.5   HIPERLAN/2 Reference Model [12].

the Physical Layer encompass data rates from 6 Mbps up to 54 Mbps with 20 MHz spacing between adjacent channels. The MAC layer still utilizes the CSMA/CA scheme.

3.2   BRAN HIPERLAN STANDARDS [4]

ETSI Project Broadband Radio Access Networks (BRAN) specifies a family of wireless network standards, which are collectively referred to as High Performance Radio Local Area Networks (HIPERLAN), to jointly support a wide variety of usage scenarios and applications. The BRAN family of standards includes HIPERLAN Type 1 (high speed wireless LANs), HIPERLAN Type 2 (short range wireless access to IP, ATM and UMTS core networks) both operating in the 5 GHz bands, HIPERACCESS (fixed wireless broadband point-to-multipoint radio access typically supporting 25 Mbps data rate) operating in various bands and HIPERLINK (wireless broadband point-to-point interconnection at very high data rates up to 155 Mbps over 150m distance) operating in the 17 GHz band. The HIPERLAN/1 standard was approved in 1996, but no products appeared in the market. The HIPERLAN/2, which is currently under development, is believed to be able to replace the old Type 1 standard. In the rest of this section, brief overview of HIPERLAN/2 is presented.

HIPERLAN type 2 is confined to the two lowest layers of the open systems interconnection (OSI) model, the physical and the data link control layer. The basic approach taken by the ETSI project BRAN is to standardize only the radio access network and some of the convergence layer functions to different core networks. The core network specific functions will be left to the corresponding forums (e.g., ATM Forum and IETF) as illustrated in Figure 9.5 [12].

HIPERLAN/2 has three basic layers, Physical layer (PHY), Data Link Control layer (DLC), and the Convergence layer (CL). In the PHY layer, the special form of multicarrier modulation scheme OFDM was selected but with different parameters compared to IEEE 802.11a. The DLC layer constitute logical link between an access point (AP) and mobile stations (MT). It consists of a set of sublayers, MAC protocol, Error Control (EC) protocol, Radio Link Control (RLC) protocol with the associated signaling entities DLC Connection Control (DCC), the Radio Resource Control (RRC) and Association Control Function(ACF). The MAC protocol is Time-Division Duplex (TDD) based dynamic Time-Division Multiple access (TDMA), i.e., the time slotted structure of the medium allows for simultaneous communication in both downlink and uplink within the same time frame that is called MAC frame in HIPERLAN/2. All data from both AP and the MTs is transmitted in dedicated time slots, except for the random access channel where contention for the time slot is allowed. The CL layer has two main functions, adapting service request from higher layers to the service provided by DLC and to convert the higher layer packets with variable (or fixed) size into fixed size that is used within DLC. There are currently two different types of CLs defined, cell-based and packet-based. The former is intended for interconnection to ATM networks while the layer can be used in a variety of configurations depending on fixed network type and how the internetworking is specified.

3.3 SHARED WIRELESS ACCESS PROTOCOL (SWAP) OF HOMERF [6,7]

The SWAP specification defines a common air interface that supports both wireless voice and LAN data services in the home environment. SWAP operates in worldwide available 2.4 GHz ISM band using digital Frequency Hopping Spread Spectrum technique. It combines elements of the existing Digital Enhanced Cordless Telecommunications (DECT) and the IEEE 802.11 standards. On the other hand, the elements from both DECT and 802.11 specifications are adapted to lower the cost of system deployment. The protocol architecture resembles the IEEE 802.11 wireless LAN standards in Physical layer and extends the MAC layer with the addition of a subset of DECT standards to provide isochronous services such as voice. The SWAP supports both a TDMA service to provide delivery of interactive voice and other time-critical services, and a CSMA/CA service for delivery of high-speed packet data, such as TCP/IP (Transmission Control Protocol and Internet Protocol).

The SWAP system can operate either as an ad hoc network or as a managed network under the control of a Connection Point. In an ad hoc network, where only data communication is supported, all stations are equal and control of the network is distributed between the stations. For time critical communication such as interactive voice, a Connection Point is required to coordinate the system. The Connection Point, which provides the gateway to the PSTN (Public Switched Telephone Network), can be connected to a PC via a standard interface such as USB (Universal Serial Bus) that will enable enhanced voice and data services. The SWAP system can also use the Connection Point to support power management for prolonged battery life by scheduling device wakeup and polling. The network can accommodate a maximum of 127 nodes of four basic types, Connection Point that supports voice and data services, voice terminal that only uses TDMA services to communicate with a base station, data node that uses the CSMA/CA service to communicate with a base station and other data nodes, and voice and data node which can use both types of services.

3.4   BLUETOOTH SPECIFICATION [8,9]

The Bluetooth Special Interest Group (SIG) has developed the Bluetooth specification that allows developing interactive services and applications over interoperable radio modules and data communication protocols. Bluetooth radios also operate in the 2.4 GHz unlicensed ISM band. A Frequency Hopping Spread Spectrum transceiver, supporting a gross data rate of 1 Mbps, is applied to combat interference and fading together with forward error correction (FEC). A shaped, binary FM modulation is applied to minimize transceiver complexity. TDD radio access scheme is used for full-duplex transmission. The Bluetooth is a combination of circuit and packet switching. Slots can be reserved for successive packets that need to be synchronized. Each packet is transmitted in a different hop frequency. A packet nominally covers a single time-slot, but can be extended to cover up to five slots. Bluetooth can support an asynchronous data channel, up to three simultaneous synchronous voice channels, or a channel that simultaneously supports asynchronous data and synchronous voice.

The Bluetooth system supports both point-to-point and point-to-multipoint connections. Several piconets can be established and linked together in an ad hoc manner, where each piconet is identified by a different frequency hopping sequence. All users participating the same piconet are synchronized to this hopping sequence. The topology can best be described as a multiple piconet structure.

4   CHALLENGES AND FUTURE TRENDS

At present, one of the main challenges encountered by the wireless data communication industry is the interoperability between various standards, i.e. IEEE 802.11, HIPERLAN/2, HomeRF SWAP and Bluetooth. The IEEE 802.11 and the HIPERLAN/2 are incompatible and competing standards applying to almost the same wireless LAN applications. But with the increasing role of the laptops carried by business people and students in retrieving information through Internet, interoperability between these two standards is greatly desired. If they are compatible, people can access Internet with laptops when one comes within a wireless LAN service (free or charged service) area in campus, office building, conference, meeting room, exhibition hall, airport, hotel, etc. The market of HomeRF partly overlaps with wireless LAN and partly with Bluetooth. As a result, the same issue of interoperability exists for HomeRF. Users would not like three different technologies, which only means they have to buy three different things to do almost the same job. In the future, the 'plain' interoperability is most probably not a limiting issue due to multimode terminals (e.g. by software radio technology) and intersystem-roaming possibility.

Another main concern arises from Bluetooth. The intended Bluetooth devices (such as cellular phone, camera, notebook and headset) are mostly carried along by users. So, there are great chances that Bluetooth devices come into the service areas of wireless LAN or HomeRF. On the other hand, Bluetooth is always on and is designed to automatically configure itself into an ad hoc network as devices come within the range of each other. HomeRF and IEEE 802.11 frequency hopping system as well as Bluetooth system all operate in the unlicensed 2.4 GHz ISM band using Frequency Hopping Spread Spectrum (FH/SS) technology. By using FH/SS, the available frequency spectrum is divided into a number of channels and the radio transmitter hops from channel to channel in a predefined sequence. If the transmission on any channel is corrupted by interference, the transmitter retransmits until possible. As long as there are enough channels and few enough transmitters, there will be no interference between coexisting FH/SS radio systems. Fast hopping Bluetooth signal is very possible to kill wireless LAN packets, which has much slower hopping rate. In the long run, Bluetooth will have to become interoperable with wireless LAN and HomeRF. As the finalization of IEEE 802.11a and HIPERLAN/2 approaches, one possible solution is that wireless LAN moves to the 5 GHz band and thus avoiding interference. Major wireless LAN suppliers such as Proxim are already considering integrating Bluetooth and wireless LAN technology in the same radio transceiver to eliminate the possibility of interference.

Integrating geolocation and context awareness is seen to be one of the major trends of wireless data communication systems [10]. Such context adaptability and awareness are relatively new tools for the designers of wireless systems and services. Geolocation information is perhaps the most useful context information for wireless terminals. By exploiting geolocation information, applications can adapt their behavior to the changes in locations and operating environments in order to improve performance or to provide new geolocation-based services.

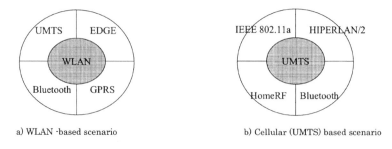

a) WLAN -based scenario   b) Cellular (UMTS) based scenario

Figure 9.6   Application scenarios for inter-technology networks.

Another area of new applications is the scenario of using WLANs as a hot spot service for cellular networks such as UMTS (Universal Mobile Telecommunication Service). Mobile access to the Internet will be made available in the future over heterogeneous wireless networks. Data services of cellular networks will be augmented by spottish high-data rate WLAN services in dynamic and seamless manner. This requires inter-technology mobility management [11], which can take place in several layers of communication. These features are also believed to be part of the so-called 4th generation wireless mobile system.

Introducing inter-technology mobility brings additional value for both cellular and WLAN networks by increasing usability and scalability. As Figure 9.6 illustrates, different application scenarios can be defined around inter-technology mobility. One is that WLAN is the primary (underlay) network (Figure 9.6a) while UMTS, EDGE (Enhanced Data for GSM Evolution, an upcoming extension to the GSM standard for higher data rates), Bluetooth, GPRS (General Packet Radio Service) networks are overlayed. For instance, a user has a laptop in his office connected to the company WLAN. When the user leaves office, he may want to maintain the network connection alive (having some application, such as ftp or newsgroup, running). Then the user can maintain the WLAN connection as long as possible and switch to the overlaying cellular data services dynamically.

Another application scenario (Figure 9.6b) is value-added cellular data servicesfor mobile users. Here the primary (underlay) network is the cellular network, which can be overlaid with spottish high data-rate WLANs in the areas such as business centers, hotels, airports and so on. It can be either provided as high-speed mobile data service with extra charge, or as a promotion to attract mobile users to visit certain commercial locations, such as shopping centers or specific airlines. Providing inter-technology mobility or roaming capability is seen to be one of the central issues in the emerging fourth generation of telecommunication networks and systems.

5   CONCLUSIONS

In this paper, we presented an overview of the emerging applications and standards of wireless data communication systems. The purpose of this paper is to provide the reader an overall view of the current status and future trends of wireless data communication technologies. The great potentials of wireless data communication systems have not been fully exploited yet. With the emergence of new applications such as HomeRF and Bluetooth, the market has been growing fast and considerable interests of both researchers and service providers have been attracted to wireless data communication areas. However, developing new applications within and beyond the scope of wireless LAN, HomeRF and Bluetooth to further expand the market still remains as a challenging task for the industry. One of the important trends of wireless networks is that WLANs, piconet and cellular networks will be all integrated with the aid of emerging multimode terminals. Thus another challenging task would be cost-efficient manufacturing of multimode terminals to make interoperability possible and to make the 4th generation possible.

Acknowledgments

The authors would like to express their appreciation to Dr. Jacques Beneat and Dr. Prashant Krishnamurthy, our colleagues at CWINS and also Dr. Roman Pichna (currently at Nokia), Dr. Jaakko Talvitie (currently at Elektrobit Ltd) and Mr. Juha-Pekka Makela at CWC in the University of Oulu, Finland, for fruitful discussions and a variety of help.

References

[1] K. Pahlavan and A. Levesque, *Wireless Information Networks*, John Wiley & Sons Inc., 1995.

[2] K. Pahlavan, A. Azhedi and P. Krishnamurthy, "Wideband Local Access: Wireless LAN and Wireless ATM", *IEEE Communication Magazine*, pp. 34-40, November 1997.

[3] Jim Geier, *Wireless LANs: Implementing Interoperable Networks*, Macmillan Technical Publishing, 1999.

[4] M. Johnsson, "HiperLAN/2 - The Broadband Radio Transmission Technology Operating in the 5 GHz Frequency Band, version 1.0", *HiperLAN/2 Global Forum:* http://www.hiperlan2.com, 1999.

[5] R. Ganesh, K. Pahlavan and Z. Zvonar, *Wireless Multimedia Network Technologies*, Kluwer Academic Publishers, 2000.

[6] HomeRF Homepage: http://www.homerf.org/.

[7] "The Shared Wireless Access Protocol: Voice & Data Communications for the home", *HomeRF SWAP White Paper:* http://www.homerf.org/, March 1998.

[8] Bluetooth Homepage: http://www.bluetooth.com/.

[9] R. Mettala, "Bluetooth Protocol Architecture, version 1.0", *Bluetooth White Paper:* http://www.bluetooth.com/, August 1999.

[10] P. Lettieri and M.B. Srivastava, "Advances in Wireless Terminals", *IEEE Personal Communications*, pp. 6-18, February 1999.

[11] M. Ylianttila, R.Pichna, J. Vallstrom, J. Makela, A. Zahedi, P. Krishnamurthy, K. Pahlavan, "Handoff Procedure for Heterogeneous Wireless Networks", *Proceedings of Globecom'99*, pp. 2783-2787, Rio De Janeiro, Brazil, December 1999.

[12] ETSI DTS/BRAN030003-1 V0.i, Broadband Radio Access Networks (BRAN), High Performance Radio Local Area Networks (HIPERLAN) Type 2, Functional Specification Data Link Control (DLC) layer Part 1 - Basic Data Transport Function.

[13] Justin Pearse, "Bluetooth explosion may hinder development", ZDNet UK, http://www.zdnet.co.uk/news/2000/1/ns-12588.html, January 2000.

Chapter 10

WIRELESS INTERNET - NETWORKING ASPECT

Li Fun Chang
AT & T Labs - Research
lifung@research.att.com

Abstract   It is envisioned that wireless and the Internet will merge in the foreseeable future. The challenge, and the race, is on to offer an Internet Protocol (IP) based, wide area, high speed, mobile, wireless packet data connectivity. Currently, cellular industry is preparing for the third generation wireless technologies and architectures to support wireless access to the Internet. In terms of networking and mobility management, two approaches have been considered. EGPRS and W-CDMA network uses cellular-like protocols for mobility management whereas cdma2000 employs Mobile Internetworking Protocol (Mobile IP) originally designed for mobility management within wireless local area network for wide area mobility management. This chapter provides overview of these two approaches, presents a generic design for wireless IP network and identifies challenges and future directions for wireless IP services.

Keywords: Mobility Management, Mobile IP, EGPRS Networks

1 INTRODUCTION

For the past few years, we have witnessed a tremendous growth rate of cellular services for the traditional voice application and explosive subscription rate of the Internet services. The popularity of www, e-commerce, has lead the cellular service providers to consider offering value-added wireless Internet and data services to boost revenue and to attract/retain subscribers. However, to offer high-speed wireless packet data services or wireless access to the Internet, the systems shall include the following capabilities that the current 2nd generation wireless networks do not have:

High bit rate transmission over wireless channels with maximum spectral efficiency. A fundamental capability is to offer enough wireless bandwidth to support Internet applications so that the perceived performance is acceptable to the users. The three key air interface technologies for IMT-2000, namely, Enhanced Data rates for GSM Evolution (EDGE), cdma2000 and Wideband CDMA(W-CDMA) with bit rate ranging from several kbps to Mbps make the future wireless Internet system one step closer to the reality.

Packet transmission control and medium access control (MAC). Since the offered services are for applications carried over the packet-based IP network, it is essential to extend packet transmission to the wireless access network to achieve maximum efficiency. EDGE, cdma2000 and W-CDMA all support packet transmission. However, the MAC design needs further enhancement to support applications that require different qualities, e.g. real time vs. delay insensitive applicaitons.

Wireless QoS support for integrated services. One key feature offered by the next generation wireless Internet services is the ability to offer services with different QoS profiles. For wire-line networks, many research works have been done on the QoS management in terms of data transfer (packet scheduling, buffering, classification) and signaling control (resource reservation, routing, etc.) mechanisms. Simple parameters such as peak rate, delay bound, throughput are used to classify the QoS classes. However, for wireless networks the Qos mechanisms and classification are more complicated than that of the wire-line networks. This is mainly due to the fact that the available bandwidth and the error rate of the wireless links are dynamically changed because of the co-channel interference, user location, traffic dynamics, etc.. Therefore, QoS mechanism that include wireless characteristics and radio resource management, wireless call admission control will need to be developed to support wireless Internet services with different QoS requirement.

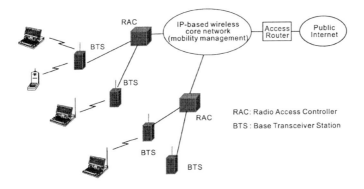

Figure 10.1 Generic Architecture for Wireless IP Network.

Voice over IP (VoIP). As the core network evolved to packet-oriented transport, it will be desirable, from operation maintenance and infrastructure deployment expenditure point of view, for the service provider to offer integrated voice and data service over a single IP-based transport. That is, to offer voice over IP (VoIP) and data services over the same wireless IP network. Call control protocol together with resource reservation protocol for wireless VoIP will need to be developed.

In addition to the capabilities mentioned above, the most important factor influencing the success of integrated wireless IP services is to offer the services at affordable price. Therefore, from service provider's point of view, it is essential to adopt an architecture that leverages the functions of the existing public IP network as much as possible so that the wireless IP network can be deployed in a cost-effective way. Currently, cellular core network architectures are migrating to packet/cell-based architecture and are all taken a similar approach by considering IP-based network as the core network as shown in Figure 10.1. In Figure 10.1, the wireless IP network consists of base transceiver stations (BTS), Radio Access Controllers (RAC or Base Station Controller BSC) and IP-based core network that support wireless specific functions such as mobility management, authentication, location management, etc. The wireless IP network is then connected to a public IP network via an access router. For instance, in GSM/EDGE/IS-136, UMTS, an IP-based Enhanced General Packet Radio Service (EGPRS) network has been considered as the wireless core network. The data packets are routed to the EGPRS backbone and then to the public Internet. While in cdma2000, the packet traffic is routed off the base station (BS) or the base station controller (BSC) via either an external or integrated packet control function (PCF) to a wireless core network. EGPRS network uses cellular-like protocols for mobility management whereas cdma2000

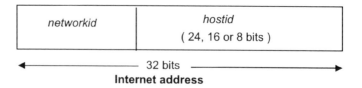

Figure 10.2   Format of Internet Address.

employs Mobile Internetworking Protocol (Mobile IP) originally designed for mobility management within wireless local area network for wireless IP services.

This chapter provides overview of the networking and mobility aspects of the wireless core networks, e.g. mobile IP-based and EGPRS-based. Specifically, we focus on the mobility protocol of the two approaches. The rest of this chapter is organized as follows. In section 2, we present a brief overview of various versions of Mobile IP protocols. In section 3, we propose a generic interim architecture for mobile access to the Internet. In this interim architecture, the 2nd generation digital voice system continues to provide voice service whereas IP-based system is employed to support packet data services. In section 4, we discuss briefly the architectures for packet data services in cdma2000. In section 5, we provide an overview of the GPRS, EGPRS and UMTS core network and associated capabilities. Finally, we conclude this chapter with future directions for the wireless IP network.

2   MOBILE IP

Existing IP routing protocols were designed for a stationary network topology. IP addresses, as shown in Figure 10.2, identify the location of an IP station (host or router) and its home network in the Internet from the network ID prefix. A protocol, Mobile IP [1], has been standardized [2] in the Internet Engineering Task Force (IETF) to allow IP stations to change their point of attachment to the network while still maintaining continuous network connectivity. That is, Mobile IP permits a mobile host to use a permanent IP address regardless of which sub-network it attaches to. It achieves this through a packet re-addressing approach, registration to the mobile agent, and encapsulation to forward data-grams to the mobile host at its current location in the network. There are several versions of Mobile IP. The RFC2002 mobile IP is normally called basic mobile IP. Other modifications such as Mobile IPv4 with route optimization and Mobile IPv6 have also been proposed to IETF to support IP host mobility. In this section, we provide a brief protocol overview of the basic mobile IPv4, mobile IPv4 with route optimization, and mobile IPv6. We also discuss commonality and differences among these protocols.

2.1 BASIC MOBILE IP

Mobile IPmobile IP architecture consists of several entities. They are mobile host (MH), correspondent host (CH), home agent (HA), home network (HN), foreign agent (FA) and foreign network (FN). A mobile host is a host that is capable of changing its point of attachment to the Internet. An MH has a permanent IP address, called the home address, which identifies the mobile's home network and does not change with the location of the station in the network. A host communicating with a MH is called a Correspondent Host (CH). An HA has a router functionality and is located in MH's home network. An HA maintains current location information for the mobile host, intercepts data-grams destined to the MH, encapsulates these data-grams and forwards the encapsulated IP packets to the MH while MH is away from its home network, and performs authentication for MH. An FA is located in MH's visited network (i.e. foreign network) and has router functionality. An FA provides routing services to the MH while registered and may serves as the default router for outgoing data-gram from MH. Home agents and foreign agents are also collectively referred to as Mobility Agents. In general, Mobile IP protocol encompasses the following basic operations: agent discovery via advertisement or solicitation, MH registration, assignment of Care of Address (COA), proxy ARP (Address Resolution Protocol) by HA, packet tunneling and triangle routing. The following lists key processes for Mobile IP:

Agent Discovery. When a mobile is away from its home network, it become aware of the Foreign Agent (FA) that serves it by exchanging messages for agent discovery with the FA. Agent discovery messages, such as agent advertisement and agent solicitation, are extensions of the Internet Control Message Protocol (ICMP) [3] router discovery messages defined for fixed hosts in the Internet. An FA transmits periodic advertisements that are broadcast or multicast to mobile stations. If a mobile station has not received agent advertisements, it can explicitly request information about the agents present in the network through agent solicitation. The mobility agents in the network that receive the solicitation reply with a uni-cast advertisement.

MH registration and COA. A home agent is made aware of the current location of the mobile stations it serves through mobile registration. Registration is required when a MH detects a change in network connectivity, the FA serving it has re-booted, or the lifetime for the current registration is nearing expiry. On receiving an agent advertisement, if the MH discovers that it is at foreign network, then it first obtains a foreign IP address called Care-of Address (COA). The COA could be associated with the IP address of the FA or a temporary address assigned to the MH via some other means, such as by a Dynamic Host Configuration Protocol (DHCP) server. The MH then updates its HA of its current mobility binding by sending a registration request packet to the HA. Note that the mobility binding kept at the HA associates the home address of the mobile to its current care of address. When the MH detects via agent advertisement message that it has roamed back to its home network, it then explicitly de-registers with its HA. Upon receiving the de-registration message, the HA deletes all bindings for the MH from its mobility binding table. While at home network, the MH behaves like a stationary IP node and uses Address Resolution Protocol (ARP), Reverse ARP (RARP) or other well-known mechanisms to make itself known to its home network.

Packet Delivery.   When a correspondent host (CH) originates a packet to a mobile host, it includes the MH's IP address as the destination address in the IP packet header. Note that the CH knows the home IP address of the MH and it need not be aware that the destination is a mobile. Since the MH's IP address is associated with the MH's home network, the packet is routed to the home network using normal IP routing protocols. When the packet reaches the home network, a router connected to the network launches an ARP request to determine the hardware address of the mobile host. When the mobile host is located in its home network, it sends and receives IP data-grams like an ordinary stationary IP node. It receives the ARP request and it responds with its link layer address in an ARP reply. The packet is then delivered to the mobile. When the MH is away from home, the HA responds to the ARP query launched by the home network router with a *proxy ARP reply* informing the router that HA will serve as a proxy for the mobile. The home network router will then delivered any packet destined to the MH to the HA. The HA encapsulates the IP packet into another IP packet with the destination address set to the MH's current care of address. At the foreign agent, packets are de-capsulated and delivered to the mobile node. This process of routing IP datagrams through the HA is called *triangular routing.* Figure 10.3 illustrates triangular routing process.

Note that packets originated by the MH are routed directly to the destination, using standard IP routing. If the FA is the mobile station's default router, IP data-grams are sent to the FA, which routes them to the destination host. The HA is not involved in data-gram delivery from the MH, unless the MH desires location privacy or firewall is implemented in the foreign network. In such a case, the MH can choose to send encapsulated IP data-grams originated by it to its HA. The HA decapsulates and delivers the packet. This process is called reverse tunneling which has also been standardized by IETF as a means to

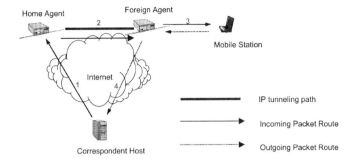

Figure 10.3   Diagram for Triangular Routing.

route MH originated packets while MH is in the foreign network that employs source address dependent routing mechanism.

2.2   MOBILE IPV4 WITH ROUTE OPTIMIZATION

In the basic mobile IP routing procedure, triangle routing and tunneling can result in possible sub-optimal routing of IP data-grams. It not only creates additional load to the home network but also introduces additional delay in data-grams delivery. The route optimization protocol [4], proposed in the IETF for IPv4, enhances the basic Mobile IP protocol by avoiding sub-optimal routing of IP data-grams through the HA. The basic concept is as follows. When the MH is away from home, the HA tunnels the packet to the MH at it's foreign location. Simultaneously, the HA sends an appropriate "Binding Update" message to the CH, this message includes MH's current COA. If the CH implements route optimization, it caches the association between the MH's home address and it's COA in its binding table together with the registration lifetime for validity of the binding duration. All future data-grams from the CH are first encapsulated by the CH and are then sent directly to the MH in its foreign network. Figure 10.4 illustrates the path for the packet delivery before and after the binding update at the CH.

The route optimization protocol also includes a process for the MH to inform the previous FA its new COA when the MH moves to a new foreign network. This is done as part of the registration process, the MH requests its new FA to notify its previous FA on its behalf by including a previous foreign agent notification extension in its registration request message. With this procedure, when a CH sends packets to the MH using obsolete COA, the previous FA will be able to tunnel these packets to the MH's new FA and thus minimized packet loss rate during MH movement. At the same time, the previous FA will send a "Binding Warning" message to the HA notifying HA to update the CH with

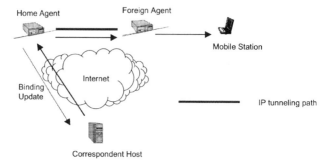

(a) Packet Delivery Before Binding Update

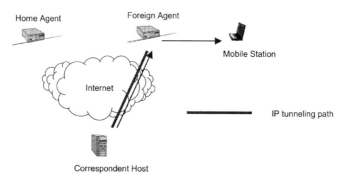

(b) Packet Delivery After Binding Update

Figure 10.4   Mobile IPv4 with Route Optimization, Before (a) and After (b) Binding Update.

MH's new COA. Consequently, an optimized route between MH and CH can be restored.

2.3   MOBILE IPV6

IPv6 [5], a new version of the Internet Protocol, has been specified by IETF as the successor to IPv4. The most notable features of the IPv6 is the increase of the IP address size from 32 bits to 128 bits and the inclusion of the routing information in the header. Mobile IPv6 [6] is the protocol specified to support mobility in the system using IPv6. The overall operation is as follows. When MH moves to a different network, MH acquires a new COA and registers this new COA with its home agent by sending a binding update message to its HA. The HA acknowledges the receiving of the binding update message by returning a "Binding Acknowledgement" message to the MH. CHs, without binding cache entry for the MH, will send the packets to the MH using its home IP address. The packet will be routed to the HA via normal IPv6 routing protocol. The HA then encapsulated the packet with MH's COA and tunnels it to the MH. The MH receives the tunneled packet from its home agent and performs de-capsulation. The MH assumes that the correspondent host does not have any binding cache entry for the MH, since otherwise the CH will send the packet directly to the MH using a Routing header in the IPv6 header extension. The MH then decides to send a binding update together with its lifetime directly to the CH for future packet delivery from CH with route optimization. A CH with MH's binding entry in the cache may send a "Binding Request" message to the MH whenever the binding is near expiration. Upon receiving the binding request message from the CH, the MH will then return a "Binding Update" message to the CH.

After creating a binding cache entry for the MH, the CH will use a Routing Header (feature for IPv6) to route the packet to the MH. That is, in the packet's IPv6 header the destination address is set to the MH's COA copied from the binding cache and the routing header is initialized to the home address of the MH. When the MH receives the packet from the CH using a routing header, the MH replaces the destination address of the receiving packet with the address (the home address of the MH) in the routing header and then passes the packet to the higher layer protocol. Figure 10.5 depicts the packet delivery for Mobile IPv6.

In summary, although basic Mobile IPv6 and Mobile IPv4 share many common features, one can observe major differences from the above description as follows:

- Route optimization is an integral part of the Mobile IPv6 and is performed together with the registration process by a single protocol rather than two separate processes as in Mobile IPv4. Moreover, binding update message does not need be sent separately from the data traffic. In the IPv6 header extension, binding update message is coded and carried by a special option in the destination header so that same packets from MH to CH, or from MH to HA can carry data traffic and binding update in the header extension.

- No FA is required to provide any special supports for the MH. The MH uses IPv6's neighbor discovery protocol or other means to acquire COA and the MH performs packet de-capsulation function.

- Packets from CH to the MH are directly routed to the MH by using feature of the IPv6 routing header instead of IPv6 encapsulation.

- Binding update is initiated by mobile node to HA or CH instead of HA.

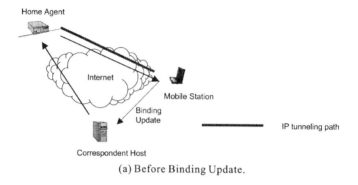

(a) Before Binding Update.

(b) After Binding Update

Figure 10.5   Packet Delivery for Mobile IPv6 Before (a) and After (b) Binding Update.

2.4   SUMMARY

The Mobile IP protocols have been designed to be independent of the access/air interface. Therefore, any future wireless Internet service providers can offer Mobile IP service by incorporating Mobile IP functions at appropriate network nodes and by designing protocols to enable inter-working between Mobile IP protocol and wireless mobility management protocols. For instance, Mobile IP Ad-Hoc group in the 3GPP (3rd Generation Partnership Project) has outlined requirements, necessary changes and interworking protocols for providing Mobile IP service in the GPRS (General Packet Radio Services) [7]. Another approach of using Mobile IP for the future wireless Internet services is to use enhanced/modified Mobile IP protocol for wide area mobility management instead of the traditional cellular mobility management approach using HLR (Home Location Register) and VLR (Visitor Location Register). In IETF, the Mobile IP Working Group is developing techniques for this purpose. Specifically, the group has identified several tasks including QoS, security, location privacy, etc. Some of the key concerns for using Mobile IP for wide area mobility management are:

1. **Agent discovery:** Mobile IP relies on agent advertisement or solicitation messages to detect the MH's movement into a foreign network. These messages are network layer messages. For wireless networks that provide routing area information or location area information in the system control channel, the agent discovery or solicitation messages are redundant and shall not be sent over the air. Consequently, when the MH performs cellular routing area update or location area update, the network needs to be able to perform mobile IP registration on behalf of the MH if necessary.

2. **Registration delay:** Mobile IP registration request and reply messages are carried over UDP and routed to the HA which may be many hops away from the visited network. The delay involved in this process may not be acceptable for some real time data services.

3. **Authentication:** the authentication of the MH is performed at the HA in the Mobile IP protocol design. This may result in significant registration delay when MH moves from one subnet to another within the same domain of the foreign network. Therefore, security association between FA and HA is required.

4. **Authorization & accounting:** Mobile IP protocol does not provide any mechanism for authorization, or accounting, thus it is crucial to include procedure either via lookup of the user profile in HLR or using AAA (Authentication, Authorization and Accounting) protocols to support inter-domain and intra-domain mobility. A comprehensive discussion on the functional and performance requirements that Mobile IP places on AAA protocols can be found in [8].

3   CELLULAR TO WIRELESS IP: AN INTERIM ARCHITECTURE

In the previous session, we have provided a comprehensive overview of Mobile IP and identified some key enhancements required for Mobile IP to provide wide area mobility management for future wireless IP services. In this session, we demonstrate the use of Mobile IP for wide-area mobility management for mobile access to the Internet assuming the required enhancements are available. We present a generic interim architecture for a wireless network that supports voice service as well as best effort data services. In this particular architecture, cellular registration and mobility management for voice services uses much of the existing cellular infrastructure that consists of the VLR, HLR,

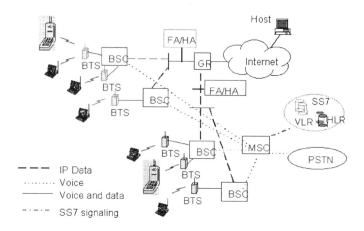

Figure 10.6   Proposed Inter-working Architecture for Cellular to the Internet.

and the AC (Authentication Center). Mobility management of users accessing the Internet is based on Mobile IP architecture and some additional capabilities of the radio system. The overall architecture is illustrated in Figure 10.6 where much of the existing cellular infrastructure used for providing voice service is utilized. The Mobile Subscriber (MS) uses the same physical channel over the air interface with the BTS for both voice and data. However, to provide more reliable and efficient transport for data traffic, a Radio Link Protocol (RLP) suitable for bursty data traffic is employed over the air interface. At the BSC, the voice traffic is routed to the MSC, while the data traffic is routed to a Gateway Router (GR), an interface node between the wireless data network and the external public Internet. The data network consists of one or more sub-networks. Each BSC is an IP node on the sub-network and has an IP address. The BSCs are capable of IP layer functions and performs routing based on MS's IP address. An MSC serves one or more BSCs. The MSC interfaces to the PSTN for routing the voice calls and to the SS7 for cellular mobility management.

Note that GR acts as a gateway between the wireless data network and the Internet. If multiple sub-nets are connected to the GR, it routes the data-grams to the appropriate sub-net. The GRs within a wireless network are interconnected to route packets between sub-networks in the wireless network. The FA and HA functions may be combined with the GR. In this scenario, when an MS is in the home sub-net, the GR routes the data-grams to the appropriate home sub-net. When the MS is visiting another sub-net, the MS registers with the FA and the HA routes the data-grams to the corresponding COA of the FA.

Another option is to have only the HA functions combined with the GR, and leave the FA functions in individual sub-nets.

Each wireless network is divided into a number of voice registration (location) areas. We assume separate mobility management for voice and data users. Therefore, the domain of a sub-net for data service and a cellular registration area are independent of each other. A data sub-network may be a subset of cellular registration area, cover multiple cellular registration areas, or may overlap between two cellular registration areas. However, the cellular registration area is the same as voice service registration area.

The mobility needs of the data users that impact the mobility management design are as follows:

- Data registration areas follow Mobile IP registration procedures and each data registration area is the domain of a data sub-net.

- A data MS may be in the Idle or Ready states. In the Idle state, the MS cannot be reached and the data calls cannot be delivered. In the Ready state, the data network is aware of the MS's location is known and the data-grams are routed to the BS serving the MS.

- When the MS is in the Ready state, that MS can be located within a cell. Therefore, as the MS moves around, the BTS that the MS is listening to will be updated using a data location update procedure.

- When in the Ready state, the MS may continuously monitor the channel for incoming packets or may follow a sleep mode. If the MS monitors the channel continuously, packets can be delivered without any prior alerts. If the MS follows a sleep mode, the MS is alerted only on the BTS serving that MS and the packets are delivered within a short period following the alert. There is no need for the MS to respond to the alert.

3.1   REGISTRATION SCENARIOS

We illustrate various registration scenarios using Figure 10.7. The figure shows two wireless networks and four data sub-nets. Wireless network 1 consists of three sub-networks, two GRs and two wireless/cellular registration areas, and wireless network 2 consists of one sub-network, one GR and one wireless/cellular registration area. Each data sub-net has an HA for data users assigned to that sub-net and an FA for data users roaming from other sub-networks. Cellular mobility functions and base transceiver stations are not shown in the figure to reduce clutter. The MS's home sub-net is sub-net 1.

The system information broadcast by the cellular network includes cellular registration area identification and BTS identification. Therefore, when an MS moves from one BTS to another, the MS can determine if it has crossed a cellular registration area or not. However, system information does not include

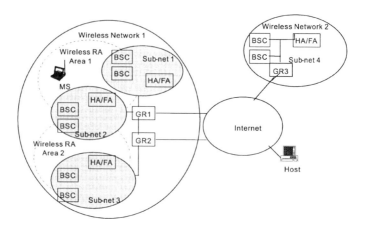

Figure 10.7   Illustration of Wireless/Cellular Registration Area and Data Sub-net.

Mobile IP registration parameters (sub-net mask, Agent Advertisement, etc.) as frequent transmission of these parameters consumes too much system information bandwidth. Mobile IP registration parameters are conveyed to MS when a BSC determines that an MS has moved from a different sub-net.

3.1.1   MS Moves within the Sub-net.   When the MS detects that it has moved from one BTS to another, it sends a Data_Location_Update to the BSC. The message includes the MS's IP address, the HA address and the COA of the serving FA. The BSC verifies if the MS has moved from a BTS in the same BSC or from a different BSC. If the MS has moved from the same BSC, the BSC updates its internal MS-BTS association. The BSC sends a Data_Location response to the MS. The BSC will continue to receive the data-grams destined for the MS and forward them to the new BS.

   If the MS moves from one BTS to another served by a different BSC, but within the same sub-net. The new BTS may belong to the same or to a different cellular registration area (e.g. sub-net 2 in Figure 10.7 where two BSCs belong to two different cellular registration areas). If the new BTS is in the same cellular registration area, then cellular registration procedure is not required; otherwise, cellular registration is performed. The MS initiates a data location update procedure with the data network. The MS sends a Data_Location Update request to the BSC. Based on the information received in the Data_Location Update request, the BSC verifies that the MS is currently registered in the same subnet. The BSC updates its internal MS-BS association and sends a Data_Location response to the MS. In order to receive the packets destined for this MS at this BSC, the BSC sends an unsolicited ARP reply to the FA to update the ARP cache binding in the FA for this MS with the physical address of the BSC. A detailed signaling procedure can be found in [9].

3.1.2   MS Moves between Sub-nets.   In this scenario, the MS has moved from one sub-net to another sub-net. The two sub-nets may belong to the same wireless registration area (e.g. sub-net 1 to sub-net 2) or to different wireless registration areas (sub-net 2 to sub-net 3). If MS detects that the new BTS is in a new cellular registration area, it performs the cellular registration. After performing the cellular registration (if required), the MS updates the data location. The MS sends a Data_Location Update request to the BSC. Based on the information received in the Data_Location Update request, the BSC determines that the MS is currently registered on a different sub-net. The BSC sends an Agent Advertisement to the MS providing new FA's source address, COA and the lifetime. Upon receiving the Agent Advertisement, the MS performs Mobile IP registration process. During the mobile IP registration process, BSC simply serves as relay point. After completing the Mobile IP registration procedure, the MS sends a Location_Update confirm message to the BSC confirming the success of the data registration procedure. The BSC sends an ARP Reply to the FA providing an association between the MS's IP address and the BSC's hardware address. The FA updates its ARP cache for the MS with the hardware address of the BSC. In order to route the IP data-grams destined for this MS to the appropriate BTS, the BSC records in its internal table the association between the MS and the serving BTS.

3.2   IP DATA-GRAM EXCHANGE

After completing cellular registration and Mobile IP registration, the MS can exchange packets with other hosts on the Internet. Exchange of packets uses Internet protocols in association with packet protocols over the air interface. The host on the Internet sends data-grams using the MS's permanent IP address. These data-grams are routed through the Internet on to the GR serving as a gateway to the wireless network. The ARP cache in the GR provides the association between the MS's IP address and the hardware address of the BSC serving the MS. The GR delivers the data-grams to the hardware address on the BSC serving the MS. The same hardware address may be serving multiple MSs and the BSC reads the IP header to determine the appropriate MS. Using the MS-BTS association table stored in the BSC, the BSC determines the BTS currently serving the MS. The BSC instructs the BTS to send the data-gram as a series of lower layer link fragments. In the other direction, the MS sends data-grams to the BSC and the BSC forwards them to the GR. The GR forwards the data-grams to the Internet and the data-grams are routed to the host.

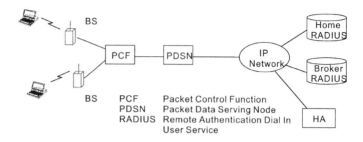

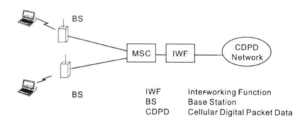

Figure 10.8   cdma2000 Packet Data Services Architectures, Type 1 (a) and Type 2 (b).

3.3   SUMMARY

The cellular voice and data network architecture proposed in this session use much of the existing cellular infrastructure. The wireless data network is modeled after IP networks and consists of several IP sub-nets. Mobility management and protocols for data network are independent of the cellular voice network. Data registration and data-gram exchanges follow Mobile IP protocol. This architecture is general and maintains a strict separation between radio subsystem and network subsystem so that the protocols can be applicable to any wireless air interface technology. The protocols for inter-working require no changes of Mobile IP and only minimal enhancements of cellular signaling. However, it should be noted that this approach simply supports best effort data services since Mobile IP protocol does not support any QoS.

4   PACKET CDMA2000 NETWORK

In cdma2000, coexistence of the circuit-switched voice network and packet data network similar to the one presented in the previous session is considered. The only difference is that the packet data network can be a CDPD based or IP based. That is, cdma2000 specifies two types of packet data services [10, 11]. Type 1 provides packet data connections based on Internet protocol stacks whereas type 2 provides packet data connections based on the CDPD protocol stacks. The reference architectures for the type 1 and 2 services are illustrated in Figure 10.8 in which the overlay circuit-switched voice network is not shown.

In the reference architecture for the type 1 service, the PCF is responsible for radio resource allocation for packet data session, link layer handoff execution, radio link setup, etc. It is connected to the packet data serving node (PDSN) which functions as an access router for the MS. Point-to-Point (PPP) link access control protocol is used to control link status and establishes link between MS and the PDSN. The system adopts Mobile IP to provide IP layer mobility management when the MS moves from one PDSN to another PDSN [12]. For authentication, authorization and accounting (AAA), the IEFT Remote Authentication Dial In User Service (RADIUS) [13] protocol is recommended.

For type 2 services, the packet data is routed off from MSC to an Interworking function and the link layer connection using PPP is established between MS and IWF for data transfer. The IWF maintains and control a packet data transmission state for each individual MS and performs packet fragmentation. This service relies on the CDPD network control protocols, authentication process and mobility management protocol such as Mobile Network Registration Protocol (MNRP) to provide mobile packet data services.

The data transmission planes for type 1 and type 2 services are shown in Figure 10.9. In either type of services, PPP is used as the link connection protocol for packet data session. Radio Link Protocol (RLP) manages point-to-point communications over the radio interface and provides reliable transmission of signaling control information, packet data information over the air link. Medium Access Control (MAC) layer consists of functions such as MAC control states, QoS control, resource allocation between competing services and competing mobile stations, and multiplexing of information from multiple services onto available physical channels.

5   GPRS (GENERAL PACKET RADIO SERVICE)/EGPRS AND UMTS NETWORKS

5.1   GPRS

In contrast to the cdma2000's approach of using CDPD or IP-based protocol as network layer protocol, GSM community has developed a new set of network layer protocols to support packet data services to the Internet or X.25 network packet data networks that is widely used in the Europe. The service is called General packet radio service (GPRS) [14, 15, 16] which has been

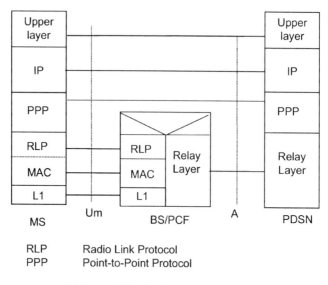

RLP   Radio Link Protocol
PPP   Point-to-Point Protocol

(a) Protocol Stacks for Type I Service.

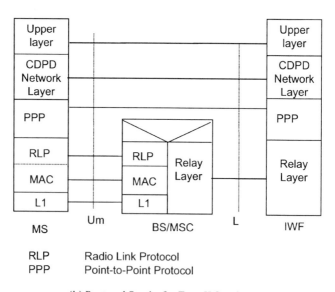

RLP   Radio Link Protocol
PPP   Point-to-Point Protocol

(b) Protocol Stacks for Type II Service.

Figure 10.9   Data Transmission Plane for Type 1 (a) and Type 2 (b) services.

introduced in the GSM phase 2 standard. The system consists of the packet wireless access network and the IP-based backbone with a special design protocol called GPRS Tunneling Protocol (GTP) to provide access to packet data networks such as X. 25, the public Internet or any future networking protocols. The GPRS packet data network includes a TDMA-based, packet switched, radio technology with 200 kHz channels, a time frame structure similar to GSM, but with four modes (GPRS) of transmission with payload bit rate ranging from 8.8 kbps to 21.4 kbps per time slot.

The logical architecture of GPRS is depicted in Figure 10.10. Two network entities are introduced in the original GSM architecture, they are the serving GPRS support Node (SGSN) and the Gateway GPRS support Node (GGSN). SGSN is at the same hierarchical level as the MSC, it keeps track of the individual MSs' location and performs security functions and access control. GGSN provides inter-working with external packet-switched networks and is connected with SGSNs via an IP-based GPRS backbone network. GGSN contains the routing information for the attached GPRS users and controls dynamic packet data protocol (PDP) address assignment if the address is not provided by the GPRS attached users during the PDP context activation process. The interface between the SGSN and the MSC/VLR is to enable MSC/VLR to send voice paging messages to the SGSN and have SGSN paged the users if users subscribe to both GPRS and GSM services. The interface between GGSN and the HLR is for the GGSN to request subscriber's location information from the HLR if needed. Thus, basically in terms of location management it still adopts the cellular VLR, HLR concept to provide wide area location management with new protocols for routing area update, cell re-selection. The Stage I of the GPRS specification is designed for best effort data services only. Currently, the standard committees are developing requirements and specifications to support multiple class QoS.

The data transmission plane used in GPRS is shown in Figure 10.11. It consists of a layered protocol structure providing user information transfer. Because GPRS is designed to support both X.25 and IP data, therefore GPRS backbone network is not fully optimized for IP. In the GPRS backbone, the GPRS Tunneling Protocol (GTP) is designed to support multi-protocol packets to be tunneled among GPRS support nodes, which are all IP-capable nodes. It is also used to carry GPRS signaling messages among GSNs. As shown in Figure 10.11, IP packets arrived at the GGSN have to be encapsulated into GTP packets, then into UDP packets and encapsulated again into IP packets for routing among GSNs. The destination GSN has to perform the reverse process to recover IP packet. Because of its flexibility for providing services to different packet data network protocols, the GTP also creates certain inefficiency in supporting pure IP services. Note that among the GSNs within the

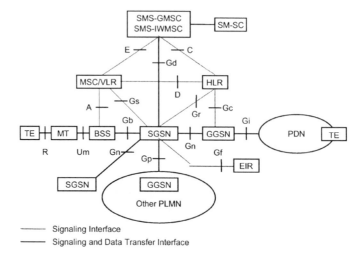

Figure 10.10   Logical Architecture of GPRS.

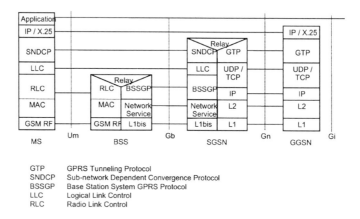

Figure 10.11   GPRS Data Transmission Plane.

GPRS backbone network, IP protocol (IPv4 or IPv6) is used for routing user data and control signaling.

For transferring the IP or X.25 packet between the serving SGSN and the mobile station, GPRS uses a different set of protocols. Subnetwork Dependent Convergence Protocol (SNDCP) maps the network layer characteristics onto specific characteristics of the underlying network. The Logical Link Control (LLC) provides a secure logical pipe between the SGSN and each mobile station. The LLC PDUs are transferred over the radio link using the services of the Radio Link Control/Media Access Control (RLC/MAC) layer. The RLC/MAC layer provides a reliable pipe responsible for transferring the LLC PDU between the Base Station System (BSS) and the mobile station. In particular, this layer is responsible for (1) mapping the LLC frames to the physical channels; (2) the access signaling and resolution procedures; and (3) providing a reliable pipe to the LLC layer for transfer of data and signaling between the SGSN and MS. The RLC/MAC layer performance determines, to a large extent, the over the air multiplexing efficiency and access delay of EGPRS applications.

The GPRS MAC/RLC layer is designed to efficiently support multiple data streams on the same Packet Data Traffic Channel (PDTCH), and to support a given data stream on multiple channels. In this sense, it facilitates the multiplexing of several bursty data transfers on the same set of physical channels. Some of the features that enable this multiplexing are discussed next.

Any data transfer in GPRS is accomplished through a *Temporary Block Flow* (TBF), which is established between an MS and the BSS and is maintained for the duration of the data transfer. A TBF is identified by a *Temporary Flow Identifier* (TFI) that is 7 bits for an uplink TBF and 5 bits for a downlink TBF. Each RLC block on the uplink or downlink has an attached TFI. The TFI is assigned by the BSS and is unique in each direction. A TBF can be *open-ended* or *close-ended*. A close-ended TBF limits the mobile station to send certain amount of data that has been negotiated between itself and its serving base during initial access. An open-ended TBF is used to transfer an arbitrary amount of data. After completion of the data transfer, the TBF is terminated and the TFI is released.

1. Downlink multiplexing of several data streams on the same PDTCH (Packet data Traffic Channel) and of a single stream on multiple PDTCHs is accomplished by assigning each data transfer a set of channels and a unique TFI. Each MS listens to its set of assigned channels and only accepts radio blocks with its TFI. Therefore, the BSS can communicate with a given MS on any of the channels assigned to the MS, and can also multiplex several TBFs destined to different MS on the same channel.

2. Uplink multiplexing is accomplished by assigning each data transfer a set of channels and a unique *Uplink State Flag* (USF) for each of these channels. Several mobiles may be assigned to the same uplink traffic channel but with different USFs. The USF is a 3 bit flag, which implies that upto eight[1] different data transfers can be multiplexed on each channel. The base uses a *centralized, in-band* polling scheme with the USF

---

[1] As USF=(000) is reserved, actually only 7 data transfers can be multiplexed on each channel.

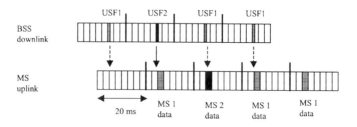

Figure 10.12   Principle of Uplink Multiplexing.

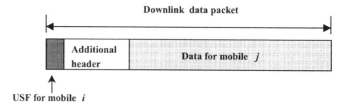

Figure 10.13   Illustration of Downlink Data Packet Structure.

flag of the corresponding downlink channel identifying the MS that is polled. Thus, an MS listens to all the *downlink* traffic channels that are paired with the set of uplink channels assigned to it: If its USF appears in the downlink channel, the MS has the right to use the corresponding uplink channel in the next frame. This mechanism is illustrated in Figure 10.12. As shown in Figure 10.13, the downlink RLC block structure enables this in-band polling. Note that even though the downlink data may be destined to one MS, the USF carried in its header can be targeted for a different MS.

To enable a packet data transfer between MS and the external packet data network, a procedure called PDP (Packet Data Protocol) context creation is used. This procedure establishes packet routing and transfer information for a particular PDP address (IP or X.25 address) in MS, SGSN and the affected GGSN. The PDP context activation process can be initiated by the MS or by the GGSN. During the MS initiated activation process, the MS informs the network (SGSN, GGSN) the type of the data services (X.25, or IP) and the associated QoS that it is requesting, and acquires the PDP address from the affected GGSN if it does not have one. The MS may also select a reference point, Access Point Name (APN), to a certain external packet data network

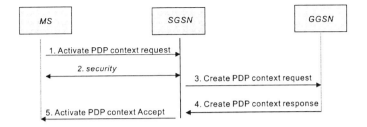

Figure 10.14 MS Initiated PDP Context Procedure for GPRS.

ɔ connect to and includes this
ɪe APN provided by the MS, if
SGSN creates a tunnel ID for
reation request to the affected
ɪn its PDP context tables and
the GGSN (1) to route packet
een the SGSN and the external
. The PDP context activation

P context is created per PDP
ɪd inefficient for many existing
ɪs. Therefore, a secondary PDP
ɪ GPRS stage 2 specifications
ɛ PDP address and other PDP
P context, but with a different
ʔ address use the same tunnel
oint identifiers to differentiate
ivation procedure is similar to
raffic Flow Template (TFT) is
ɪ. TFT is sent from the MS and
to enable packet classification

:ation process, the MS includes
e of the information elements
ɟ message. The stage I GPRS
t class of QoS and hence does
mply provides best effort data
) is now under development to
-section describes EGPRS, its

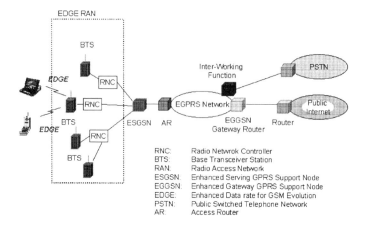

Figure 10.15   Architecture of EGPRS.

5.2   EGPRS

EGPRS is a derivative of GPRS. The key differences between EGPRS and GPRS include air interface design as well as core network capability. GPRS air interface supports transmission rate from 8.8 kbps to 22 kbps whereas EGPRS uses EDGE (Enhanced Data rate for GSM Evolution) [18] as the air interface and supports transmission rate ranging from 8.8 kbps to 59.2 kbps. Similar to GPRS air interface, EDGE is an evolution of GSM air interface. That is, it is TDMA-based, packet switched, radio technology with 200 K Hz channels, a time frame structure similar to GSM, but nine modes of transmission. The system is designed to operate in most of the current 2G spectrum allocations. Currently, the standards bodies are working on specifications for EGPRS to support integrated voice and data services. In this sub-session, we simply present its key features and capabilities.

EGPRS is designed to support integrated real-time applications and packet data over a common IP platform. In particular, real time voice traffic will be transferred into IP packets and carried over the IP-based backbone. This eliminates the need for a separate circuit switched network. The overall architecture is shown in Figure 10.15 where a new network entity, Radio Network Controller (RNC), is introduced to perform radio related functions such as handoff, radio resource management, radio admission control, etc. It is hoped that the EGPRS core network entities, ESGSN, EGGSN, will share as many commonality as the core network entities, 3G-SGSN and 3G-GGSN, of the UMTS.

As mentioned earlier, EGPRS will provide multiple classes of QoS services. In particular, four classes of services that are the same as those defined in

Table 10.1   Four traffic classes that will be supported in the EGPRS.

Traffic class	Conversational	Streaming	Interactive	Background
Fundamental characteristics	-Preserve time relation between information entities of the stream -Conversational pattern (stringent and low delay)	-Preserve time relation between information entities of the stream	-Request response pattern -Preserve payload content	-Destination is not expecting the data within a certain time -Preserve payload content
Applications	voice	streaming video	Web browsing	Background download of emails

UMTS[19] will be supported. The characteristics of the four QoS classes are summarized in Table 10.1. They range from the conversational class, e.g voice, which imposes a very stringent delay requirement, to the background class, e.g. best-effort data, which imposes a relatively loose delay requirement. The requirement to support a voice service also implies strict error rate requirements, which result from the inability to rely on re-transmissions to improve performance.

In contrast, the current GPRS specifications are designed primarily for best-effort data services with some Quality of Service (QoS) classes, all of which have loose delay constraints. For real time applications in EGPRS, the MAC/RLC designed for the GPRS is no longer applicable due to the relative long MAC delay. Thus in EGPRS, four multiplexing options are considered for voice applications. They are: static time slot allocation to a voice call without any multiplexing, static time slot allocation to a voice call and multiplexing of best effort data from the same user, static time allocation to a voice call and multiplexing of best effort data from different users, and full multiplexing with other voice calls and data. In order to support the third and fourth options, i.e. multiplexing a range of services defined for the EGPRS in a time-multiplexed manner on the same channel(s), several new capabilities or enhancements are needed. Some of these are:

1. Fast uplink access capability: This and the next capability are driven by the need to quickly assign resources when an interactive service transitions from an inactive to an active period during an ongoing session (e.g. the start of a talk-spurt in a voice conversation).

2. Fast resource assignment capability for both uplink and downlink.

3. Service differentiation at the base.

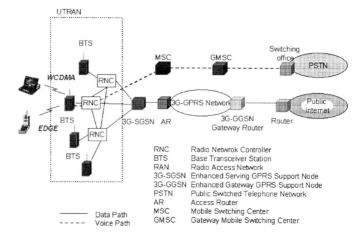

Figure 10.16   Architecture of the UMTS Network.

4. End-to-End QoS guarantees

5. Fast cell re-selection and fast handoff: This identifies a weakness in the current GPRS specifications, where an active data transfer is suspended while a terminal moves to a new base.

6. Optimal coding and interleaving for some applications: While designing mechanisms targeted for each possible application defeats the purpose of an integrated solution, it is nevertheless useful to design specific solutions targeted for a selected set of applications, e.g. voice or streaming data.

7. Header compression/elimination: 40 byte (or larger) IP headers can be an unnecessary overhead for applications with periodic, small data units.

Currently, in ETSI SMG2 EDGE workshop, efforts to support the above-mentioned enhancements and capabilities are under going. Furthermore, in 3G.IP, a consortium to develop protocols, enhancements required for EGPRS core network for providing integrated real time voice and data services, are working on service definitions, QoS requirements, call control signaling protocol for wireless voice over IP services, and wireless QoS management mechanism for the EGPRS network.

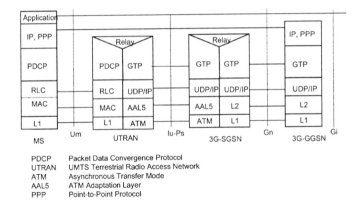

PDCP	Packet Data Convergence Protocol
UTRAN	UMTS Terrestrial Radio Access Network
ATM	Asynchronous Transfer Mode
AAL5	ATM Adaptation Layer
PPP	Point-to-Point Protocol

Figure 10.17   UMTS Data Transmission Plane.

5.3   UMTS NETWORK ARCHITECTURE

UMTS (Universal Mobile Telecommunications Systems) is the 3rd generation system under standardization by ETSI which provide services for voice as well as for packet data. The air interface for the UMTS Terrestrial Radio Access Network (UTRAN) is W-CDMA. The overall architecture is illustrated in Figure 10.16. The voice service is supported by the circuit network consists of MSC and GMSC which is similar to the 2nd generation cellular network. Packet data services, on the other hand, are provided by the network architecture similar to the core network of EGPRS except the interface between RNC and 3G-SGSN is quite different from that of the GPRS which will be discussed latter. The radio network controller (RNC) element provides overall management of UTRAN resources and interacts with the core network via signaling messages to provide service (circuit oriented & packet oriented) to subscribers. RNC performs radio connection control and soft hand-over control. Radio connection control includes service multiplexing, link layer QoS control, mapping logical service channels onto W-CDMA code channels, etc.

Figure 10.17 illustrates the protocol stack of the data transmission plane of UMTS[17]. The GTP has been extended to the RNC in the UTRAN, that is, the packet data are tunneled from the 3G-GGSN to the RNC instead of the SGSN done in GPRS. For transferring the data packet between the RNC and the mobile station, UMTS uses a different set of protocols. Packet Data Convergence Protocol (PDCP) maps the network layer characteristics onto specific characteristics of the underlying radio-interface protocol. Furthermore, it provides protocol transparency for higher layer protocols. Note that in the current phase of the UMTS standard, ATM transport is selected for the Iub (interface between BTS), Iur (interface between RNCs), and Iu-Ps interfaces. Further-more, the AAL2 has been selected as the preferred ATM adaptation layer for voice transmission (partially filled ATM cells are allowed for the low radio frame rate) and AAL5 has been selected for data transmission.

UMTS uses much of the existing GPRS packet data protocols with some modifications and enhancements for supporting multiple QoS profiles. For ex-ample, during the PDP context creation procedure as shown in Figure 10.14, after completing the second step (security functions), the 3G-SGSN performs the radio access bearer setup procedure. That is, the 3G-SGSN sends a "Ra-dio Access Bearer Request" message together with the requested QoS profile to the RNC. Upon receiving the request, RNC will execute radio admission and resource allocation processes based on the requested QoS. Once the radio access bearer is setup, the 3G-SGSN proceeds to complete the PDP context creation process as illustrated in Figure 10.14. Another example is for location management. In UMTS since soft handoff is performed at the RNC, therefore, it is important to have an efficient routing for packet transfer. Thus for loca-tion management, a routing area update procedure similar to the one used for GPRS and a serving RNC relocation procedure are used. The routing area up-date procedure moves the data path connection from the old SGSN to the new SGSN whereas the serving RNC relocation procedure completes the connec-tions between the new SGSN and the target serving RNC connected to it.

In summary, the UMTS core network architecture, associated functions and protocols are quite similar to those of the EGPRS core network. In fact, it is the goal of EGPRS to design the interface between EDGE RNC and the ESGSN appropriately so that UTRAN and EDGE RAN can share a common core network.

6   CONCLUSIONS

In this chapter, we presented several core network approaches for the wire-less IP network. In general, two classes of the network can be identified. One uses IP-based protocol (i.e. Mobile IP) for wide area mobility management such as cdma2000. Another creates its own packet data network protocols and uses much of the cellular mobility management concept for wide area mobility management. EGPRS, UMTS are in this category. Although the approaches are quite different, the common goal is to provide efficient wireless IP services with quality. The use of mobile IP for mobility management is appealing in turns of consistency and leveraging of IP-networking protocol. However, as discussed in section 2, mobile IP protocol itself has many deficiencies and re-quires many enhancements and modifications before it can be truly used for mobility management in wireless IP networks. In contrast, the approach taken by EGPRS and UMTS create its own tunneling protocol and session management protocol to allow flexibility for interworking with IP, X.25 and other packet data protocols, yet to be invented. If IP applications are the key and enabling applications for the wireless packet data services, that is, Internet is the only public data network that the wireless access will connect to, then one should consider a solution that simplifies GTP. In summary, for either approach the mechanisms to carry out QoS services remain to be a major task to be completed. The dynamic of the available radio resource, link impairments, the impact of the bursty traffic characteristics on the radio resource allocation algorithms, etc, present enough challenges on the wireless admission control and QoS support and open a rich area for future research.

References

[1] C. E. Perkins, *Mobile IP - Design Principles and Practices*, Addison-Wesley, 1998.

[2] C.E. Perkins, Editor, "IP Mobility Support," RFC 2002, IETF Network working group, October 1996.

[3] S. Deering, "ICMP Router Discovery Messages", RFC 1256, IETF, September 1991.

[4] C. Perkins, D. B. Johnson, "Route Optimization in Mobile IP," Internet-Draft, IETF Mobile IP Working Group, Feb. 1999.

[5] S. Deering, R. Hinden, "Internet Protocol, Version 6 (Ipv6) Specification," RFC 2460, IETF Network Working Group, December 1998.

[6] D. Johnson, C. Perkins, "Mobility Support in IPv6," Internet-Draft, IETF Mobile IP Working Group, October 22, 1999.

[7] 3G TR 23.923, version 0.8.0, Combined GSM and Mobile IP Mobility Handling in UMTS IP CN, 3GPP Technical Specification, 1999.

[8] S. Glass, T. Hiller, S. Jacobs and C. Perkins, "Mobile IP Authentication, Authorization, and Accounting Requirements," draft-ietf-mobileip-aaa-reqs-01.txt, Jan. 2000.

[9] V. J. Varma, L. F. Chang, "PCS-to-Mobile IP Interworking," IEEE PIMRC'99 conference proceedings, Sep. 1999, Osaka, Japan.

[10] TIA/EIA/IS-707-A-1.11, "Data Service Options for Spread Spectrum Systems: cdma2000 High Speed Packet Data Service Option 34," December, 1999.

[11] TIA/EIA/IS-707-A-1.12, "Data Service Options for Spread Spectrum Systems: cdma2000 High Speed Packet Data Service Option 33," December, 1999.

[12] 3GPP2 P.S0001, version 1.0, Wireless IP Network Standard, December, 1999.

[13] C. Rigney, A. Rubens, W. Simpson, S. Willens, "Remote Authentication Dial In User Service," IETF RFC 2138.

[14] G. Brasche et. al. "Concepts, Services, and Protocols of the New GSM Phase 2+ General Packet Radio Service," IEEE Commun. Magazine, August 1997.

[15] GSM 03 64 v6.0.0, Digital Cellular Telecommunications System (Phase 2+); General Packet Radio Service (GPRS); Overall Description of the GPRS Radio Interface, 1998.

[16] GSM 03 60 v5.0.0, Digital Cellular Telecommunications System (Phase 2+); General Packet Radio Service (GPRS); Service Description, 1997.

[17] 3GPP TS 23.060 v.3.1.0, Digital Cellular Telecommunications System (Phase 2+); General Packet Radio Service (GPRS); Service Description; Stage 2, October 1999.

[18] A. Furuskar, S. Mazur, F. Muller and H. Olofsson, "EDGE: Enhanced Data Rates for GSM and TDMA/136 Evolution," IEEE Personal Communications Magazine, June 1999, pp. 56-66.

[19] 3GPP TS 23.107 v. 3.1.1, Technical Specification Group Services and System Aspects; QoS Concept and Architecture, Feb. 2000.

Chapter 11

DIGITAL TERRESTRIAL TV BROADCASTING SYSTEMS

Makoto Itami
Science University of Tokyo
itami@te.noda.sut.ac.jp

Abstract   Recently, investigations and implementations of digital terrestrial television broad casting systems are widely performed in the world towards the full digital television broadcasting services. Realization of digital terrestrial television systems will not only provide higher quality video and audio broadcasting than conventional analogue broadcasting, but also it is expected to provide new services such as multi-channel broadcasting, data broadcasting, interactive broadcasting, etc. At present, three principal digital terrestrial broadcasting systems are proposed in Europe, USA and Japan, respectively. In this chapter, the outlines of the specifications and functions of these three digital terrestrial broadcasting systems are given and compared.

Keywords: DVB-T, MOTIVATE, ATSC, ISDB-T, OFDM, 8VSB

1   INTRODUCTION

Investigations of digital terrestrial television broadcasting systems to provide higher quality programs and advanced broadcasting services were started in USA and EU and a little later Japan also stated investigations. As the results of investigations, three different digital terrestrial television broadcasting systems are standardized. The adopted modulation scheme of each standard is classified into two modulation schemes. One is OFDM (Orthogonal Frequency Division Multiplexing) adopted by EU and Japan and the other is 8VSB (Vestigial Side Band). The System parameters and the manners of operation are very different between EU standard and Japanese standard though EU and Japan adopt the same modulation scheme, OFDM.

At present, services of digital terrestrial television broadcasting has already started and are gradually penetrating in EU an USA. In Japan, pilot tests of digital terrestrial television broadcasting and developments of receivers are being performed in the many areas towards the beginning of the full service in 2003. Other countries except EU, USA and Japan are investigating the adoption of one of these three standards and several countries have already decided to adopt one of these standards.

In this chapter, these three standards (DVB-T, ATSC standard, ISDB-T) are described and their features are compared.

2   DVB-T

The standard of digital terrestrial television broadcasting in EU is called DVB-T (Digital Video Broadcasting – Terrestrial). DVB-T is standardized by DVB (Digital Video Broadcasting) [1, 2], an institution of standardization in EU. In the DVB-T standard, OFDM is used for a modulation scheme. OFDM is a digital modulation that uses many orthogonal digital modulated carriers for digital data transmission and it makes possible to utilize frequency band very efficiently. Since the OFDM symbol length is very long, it is less affected under multipath channel and it is possible to protect data symbols from inter-symbol interference by adding a guard interval without much loss of data transfer speed.

The concept of OFDM had already been proposed in 1950s, however, it was difficult to realize practical systems for its complexity. However, the practical implementation became possible after the OFDM modulation technique using DFT (Discrete Fourie Transform) had been proposed. The first practical OFDM system is DAB (Digital Audio Broadcasting) [1] developed in EU and many interests were paid to this system. Following to DAB system, EU adopted OFDM for digital terrestrial television broadcasting standard. In Figure 11.1, the simplified block diagram of OFDM transmission system is shown [1].

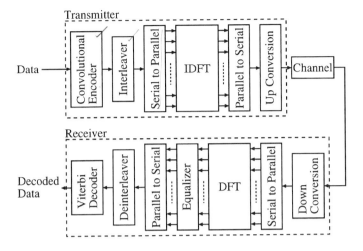

Figure 11.1   OFDM Transmission System.

By positively utilizing OFDM characteristic that is robust against multipath channels, it is possible to construct a different broadcasting network from the conventional broadcasting network, MFN (Multi Frequency Network). This new broadcasting network is called SFN (Single Frequency Network). In SFN, all broadcasting stations in the network broadcast same program at the same time as shown in Figure 11.2. Under this situation, ghost will be generated in the overlapped part of several stations' service areas and reception quality much degrades in MFN. However, OFDM is robust against multipath channel by adding a guard interval. Consequently, high quality reception in the overlapped service areas is made possible. Therefore, efficient use of frequency channel is expected in SFN. This makes easier to start new broadcasting services such as multi channel broadcasting, data broadcasting, etc. In DVB-T , SFN is expected to be positively used under many situations.

The parameters of DVB-T are shown in Table 11.1 In DVB-T standard, bandwidth of OFDM signal is about 7.6 MHz and the DVB-T standard has two transmission modes. One is the mode where the number of carriers is 1705 and the other is the mode where the number of carriers is 6817. These modes are called 2k mode and 8k mode, respectively. Each mode is named after the window size of DFT used in an OFDM modulator. As shown in Table 11.1, FEC is combined with the OFDM modulator in DVB-T to improve the reliability of the system. This is called coded OFDM (COFDM). Standard Reed-Solomon code is used as an outer code and convolutional code is used as an inner code. Interleavers are also used to improve error correction capability. In the DVB-T standard, symbol modulation scheme, guard interval

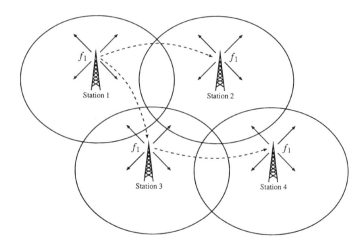

Figure 11.2   Single Frequency Network (SFN).

Table 11.1   Parameters of DVB-T.

Parameter mode	2k	8k
Band width		7.61 MHz
Number of Carriers	1705	6817
Modulation Scheme		QPSK, 16QAM, 64QAM
Effective Symbol Duration	224 μs	896 μs
Guard interval duration		1/4,1/8,1/16,1/32 of effective symbol duration
FEC (Inner Code)		Convolutional Code (1/2, 2/3, 3/4, 5/6, 7/8)
FEC (Outer Code)		Reed Solomon Code (204, 188)
Bit rate		4.98~31.67 Mbps
Video Format		MPEG-2 Video
Audio Format		MPEG-2 Audio (BC)

length, code rate of an inner code can be arbitrary selected. By various combinations of these parameters, various data transmission speeds from 4.98 Mbps to 31.67 Mbps are avaiable. The performance of system is affected by multipath and additive noise and in general these affections becomes larger as the bit rate becomes larger.

Scattered pilot symbols shown in Figure 11.3 are inserted among data symbols in order to compensate the affection of symbol distortion under multipath

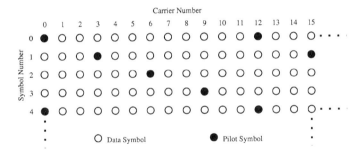

Figure 11.3   Scattered Pilot Symbol.

channel. Scattered pilot symbols are criteria used in the euqalizer for the re-cevied data symbols. In Figure 11.3, each solid circle corresponds to each pilot symbol and the other circles are data symbols. Channel transfer function against each pilot symbol can be easily derived and channel transfer functions against data symbols are estimated by interpolation using channel transfer functions against pilot symbols.

In DVB-T, various broadcasting services are possible by changing parameters. DVB-T permits multiplexing of several different programs into an OFDM signal within 7.6 MHz bandwidth. The kinds of programs that can be multiplexed are SDTV (Standard Definition Television), HDTV (High Definition Television), LDTV (Low Definition Television), Data Channel, etc.

For example, one HDTV program whose bit rate is 27.14 Mbps can be transmitted in 7.6 MHz bandwidth. SDTV signal whose bit rate is 5.75 MHz can be multiplexed up to 4 programs within the same bandwidth. SDTV program whose bit rate 5.75 Mbps and LDTV program whose bit rate is 3.38 MHz can be multiplexed up to 6 programs within the same bandwidth. Moreover, audio and data broadcasting program can be also multiplexed. By using multiplexing, multi channel broadcasting is easily realized. Consequently, various services are expected. To realize multiplexing, program provider that offers programs, multiplex provider that administrate multiplexing and network provider that manages network are necessary.

In the DVB-T standard, various broadcasting services are available and each country in EU starts some of these broadcasting services according to their requirements. The form of services will be different in each country. For example, MFN service is adopted in UK and 2k mode is used to perform this service. The reason why MFN system is used in UK is that the requirement for local broadcasting is extensive. In Germany, both SFN and MFN services are used and 2k mode and 8k mode are used. In many other EU countries, they attach great importance to SFN ans 8k mode is adopted.

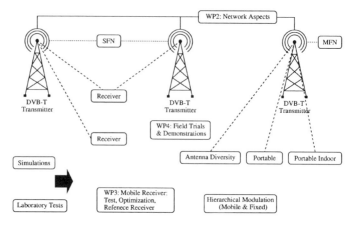

Figure 11.4   MOTIVATE Overview.

DVB-T is adopted as the standard in Australia, India, Singapore, etc.

2.1   MOTIVATE

At present, the next step of DVB-T called MOTIVATE (Mobile Television and Innovative Receivers) project is under investigation [4, 5, 6]. The MOTIVATE project investigates mobile reception techniques for DVB-T under SFN and MFN environments. It is expected that a new multimedia broadcasting network that includes mobile reception will be built under DVB-T by the MOTIVATE project. The overview of the MOTIVATE project is shown in Figure 11.4. MOTIVATE integrates mobile reception by automobiles under SFN environment, portable reception and indoor portable reception with usual stationary reception and it provides interfaces between users and information society. Here, DVB-T provides independent standard for data and multimedia broadcasting.

The followings are the main objectives of the MOTIVATE project.

- Analyze the theoretical performance limits of DVB-T for mobile reception and implement optimal receivers.

- Study, implement and test efficient algorithms for mobile SFN reception.

- Test state-of-the-art DVB-T receivers for mobile reception.

- Set-up a pilot network to measure mobile channel characteristics and mobile coverage in urban and sub urban networks.

Table 11.2 C/N ratio vs. Speed (n.w. = not working).

FFT Length	Modu-lation	Code Rate	Data Rate [Mbit/s]	Average maximum speed (channel 43 = 650 MHz)			C/N Performance at 100 km/h (* 50 km/h)		
				easy	regular	difficult	easy	regular	difficult
2k	QPSK	1/2	5	400	330	170	9	17	23
2k	16QAM	1/2	10	250	240	90	15	23	29
2k	64QAM	1/2	15	190	120	40	21	29	n.w.
8k	QPSK	1/2	5	100	70	30	10*	19*	n.w.
8k	16QAM	1/2	10	70	30	20	17*	n.w.	n.w.
8k	64QAM	1/2	15	50	25	10	26*	n.w.	n.w.

■ Set-up and carry out major demonstrations to present DVB-T in major national and international events.

■ Support integration, promotion and dissemination of results of other ACTS projects working on DVB-T.

■ Verify the open API for DVB-T receivers.

■ Give implementation guidelines for the realization of a mobile DVB-T service.

The behavior and limits of the DVB-T specification in a mobile environment are being analyzed through theoretical investigation, computer simulations, laboratory tests and field trials by the MOTIVATE project. The laboratory tests of mobile reception of DVB-T were organized in 1998 to compare the behavior of receivers and to study performance in a mobile environment. The results of measurements in several modes for 9 receivers are shown in Table 11.2. The reachable speed on average of all receivers and the necessary C/N ratio at speed of 100 km/h of 2k and 8k modes for three different channel situations are listed in Table 11.2. It is shown that 2k modes are usable in the whole UHF band and the VHF band is better suited for the 8k modes.

The field tests were performed by many MOTIVATE partners. The main results of field trials are as following.

■ Mobile applications of DVB-T are feasible using code rate 1/2 modes of the specification.

■ A data rate up to 15 Mbps using one 8 MHz channel is possible with the 64-QAM mode for mobile reception.

■ A lot of tests have shown that the field strength is the critical factor to have a good mobile reception.

- The speed of the car dit not give any restriction of mobile reception in the case of 2k mode.

- The usage of hierarchical modes of DVB-T specification is an efficient way for broadcasters to offer mobile reception and stationary reception in the same channel with different protected data streams (high priority and low priority stream).

The aim of the MOTIVATE project is to optimize next generation mobile receivers. Requirements for next generation receivers are as follows.

- Optimization of channel estimation and synchronization algorithms for mobile DVB-T

- First results using Wiener filter algorithms and a new FFT leakage equalization.

- Car PC receivers and portables.

- Receivers with diversity.

Recently mobility has become more and more important for individual and business users in the world. Currently, over 470 million people are using cellular telephones in the world and demands for mobile services are increasing. Mobile DVB-T services will complement some of these services. Mobile DVB-T services are shown in the followings. This is designed for business and individual use.

- Digital Television in cars, buses and trains.

    - EPG, traffic, navigation, weather, etc.

- Mobile contribution links.

    - MObile DVB transmission from reporting vehicles.

- Store and forward services.

    - Overnight delivery of software, movies, etc.
    - Electronic newspapers

- Mobile data broadcast.

- Individual MOTIVATE services.

    - Individually addressable applications

Table 11.3   Parameters of ATSC Standard.

Modulation Scheme	8VSB
Bandwidth	6 MHz
FEC (Inner Code)	Trellis Code (2/3)
FEC (Outer Code)	Reed Solomon Code (207,187)
Data rate	19.28 Mbps
Video Format	MPEG-2 Video
Audio Format	Dolby AC3

3   ATSC STANDARD

USA started an investigation for the next generation digital television broadcasting called ATV (Advanced Television) by ATSC (Advanced Television Systems Committee) in 1987. After several systems are proposed and baked off, the standard system using 8VSB (Vestigial Side Band) is defined. The parameters of the ATSC standard is shown in Table 11.3 [2, 3, 4]. VSB modulation is also used as in the case of conventional NTSC analogue television systems. In the ATSC Standard, by appling a VSB modulation scheme to the digital modulation, an efficient use of frequency is realized.

The ATSC standard delivers many digital terrestrial services (Video, Audio, Data, etc.). This standard is designed to meet several criteria in the followings.

- Obtain maximum coverage area by using advanced signal processing and FEC techniques.

- Deliver HDTV video data rate (19 Mbps) within existing 6 MHz channels.

- Provide robust reception under widely varying conditions (impulse, phase and thermal noise) and terrain (stationary and moving multipath).

- Have high immunity to NTSC co-channel interference.

- Be able to be constructed with low cost consumer type components.

The primary aim of the ATSC standard is to improve the quality of television broadcasting by performing digital terrestrial HDTV broadcasting. This is quite different from the aim of DVB-T. The primary aim of DVB-T is to increase available number of television channels. Therefore, an application of SFN is not considered in ATSC standard. Moreover, no special attentions is paid against mobile reception in the ATSC standard.

The block diagram of VSB transmitter is shown in Figure 11.5. In the trans-

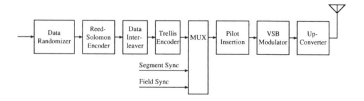

Figure 11.5   VSB Transmitter.

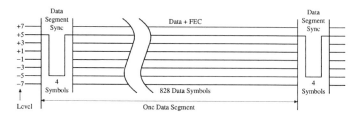

Figure 11.6   VSB Data Segment.

mitter, source data is encoded by Reed-Solomon Encoder after data random-ization. Reed-Solomon encoded data is interleaved and trellis encoded to generate 8 levels data symbols for VSB modulation using standard 4-state optimal Ungeboeck Codes. After the data symbols are multiplexed with segment sync and field sync, pilot for carrier recovery is inserted. Then VSB modulation is performed.

The VSB data segment consists of 828 data symbols (contain FEC) and the data segment sync symbols as shown in Figure 11.6. The data segment sync is used for robust symbol timing recovery and its length is that of four data symbols. Length of one data segment is 77.3 μs and it is possible to transmit 188 byte MPEG-2 data packet. Carrier pilot is created by a 1.25 level shift of VSB data segment. Symbol data is vestigial Nyquist filtered to $\alpha=11.5\%$.

By collecting 313 successive segment, the VSB field is constructed. The first segment in each field is a special segment called a data field sync. The data field sync is used for equalizer training and VSB mode identification. The structure of VSB field sync is shown in Figure 11.7. The first 700 symbols in the VSB field sync are used for equalizer training and next 24 symbols are used for VSB mode identification and next 92 symbols are reserved. Training sequence is repeated about 41 times a second.

The VSB data frame consists of two VSB data field as shown in Figure 11.8. The VSB data frame length is 48.4 μs and final symbol rate is 10.762 Msymbols/s.

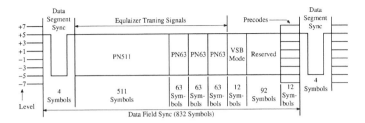

Figure 11.7   VSB Field Sync.

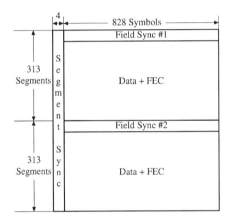

Figure 11.8   VSB Data Frame.

An ATSC standard receiver is required to meet the above specifications. The basic ATSC standard receiver features are as follows.

- AGC performed by average signal power.

- Carrier Recovery based on pilot tone.

- Timing Recovery based on segment sync.

- Equalizer training based on:

  — Training Sequence

  — Blind adaptation algorithms

  — Decision directed

- Decision directed AGC.

- Decision directed phase noise tracking.

One of the most important part of the ATSC standard receiver is the equalizer. A decision feedback equalizer is typically used and typical tap length of the feedforward part is 64 taps and that of the feedback part is 200 taps. In this specification, the feedforward part allows $\pm 3\mu s$ ghost correction and feedback part allows 19 μs ghost correction.

For equalizer training algorithm, LMS (Least Mean Square) algorithm is used for its simplicity. However, convergence speed of LMS is slow and highly depends on channel conditions. To improve this, RLS algorithm can be used. However, number of multiplication increases and it becomes difficult to make a low cost receiver.

The service of the ATSC standard is started in 1998 and about 140 stations are broadcasting in 1999 in the whole USA. This means that 70% of all viewers in USA will have at least 3 DTV signals. In the future, all commercial stations will change to DTV by 2002 and all noncommercial stations will change to DTV by 2003. Analogue NTSC broadcast is scheduled to stop in 2006.

Currently, coverage of ATSC standard is over 50% of whole USA. However, HDTV broadcasting is limited to several programs such as sports, disney movies, etc. and almost all programs are broadcasted by converting an NTSC program to HDTV format. In the ATSC standard, 18 formats that support from SDTV to HDTV are supported and each program provider can select any formats.

The ATSC standard is adopted in many countries such as Argentina, Canada, Taiwan, South Korea, etc. And South Africa, China, Hong Kong, etc. are positively investigating its adoption.

4   ISDB-T

In Japan, a report of digital terrestrial television broadcasting was submitted in 1994 a little later than EU and USA. The Japanese standard for digital terrestrial television broadcasting is called ISDB-T (Integrated Services Digital Broadcasting – Terrestrial) and was standardized in 1999. The broadcasting services are scheduled to start in 2003.

The aims of digital terrestrial television broadcasting in Japan are as follows.

- Complete transition from analogue broadcasting.

- Make the schedule of introduction clear.

- As a general rule, participation of new providers is permitted after finishing analogue broadcasting.

- Should play a role as the key broadcasting media.

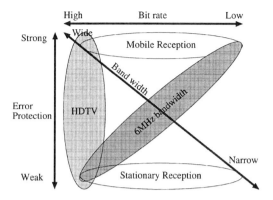

Figure 11.9   Requirements of ISDB.

Present analogue broadcasting will stop by about 2010. However, this schedule will be reconsidered in each broadcasting area under the following conditions.

- Penetration rate of the digital receivers is over 85 %.

- Should cover whole service area of analogue broadcasting in the same area.

The requirements to ISDB-T are as follows.

- Should match with other digital medias in Japan and should be extensible in the future.

- Broadcasting of one HDTV channel or three SDTV channels is possible within the present 6 MHz bandwidth.

- Mobile reception should be possible.

- SFN should be possible.

These requirements are figured out in Figure 11.9.

In order to provide various broadcasting services, Integrated Service Digital Broadcasting (ISDB) is a platform in Japan [4][7]. ISDB platform consists of three major parts; satellite digital broadcasting (ISDB-S), cable digital broadcasting (ISDB-C), terrestrial digital broadcasting (ISDB-T). Among these systems, Video and Audio and Data is integrated over standard MPEG format. And signals are transmitted MPEG-2 Transport stream packet. The concepts of ISDB is shown in Figure 11.10.

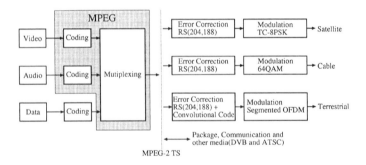

Figure 11.10 Concepts of ISDB.

In 2000, ISDB-S system using TC-8PSK starts broadcasting service and provide multichannel HDTV and SDTV and audio and data broadcasting. In this system, electronic program guide (EPG) is also provided and user can easily select preferred program.

ISDB-T adopted OFDM modulation like to DVB-T in order to satisfy the above requirements. Moreover, ISDB-T has unique features different from DVB-T. The features of ISDB-T are as follows.

- Three modes used in mobile reception, stationary reception, SFN reception are provided.

- DQPSK can be used for the modulation scheme under mobile reception.

- Adoption of time interleaving for mobile reception.

- Adoption of BST-OFDM (Band Segmented OFDM).

The parameters of ISDB-T are shown in Table 11.4. The bandwidth of ISDB-T is about 6 MHz and three modes (Mode 1, Mode 2, Mode 3) are provided according to the number of carriers. The number of carriers in Mode 1 is 1405 and this corresponds to 2k mode in DVB-T. The number of carriers in Mode 3 is 4992 and this corresponds to 8k mode in DVB-T. In ISDB-T, one more mode where the number of carriers is 2809 is defined. This mode is called Mode 2. In Mode 2, parameters have intermediate values between Mode 1 and Mode 3. Mode 2 is expected to be suitable for mobile reception under SFN because Mode 1 is suitable for mobile reception and Mode 3 is suitable for stationary reception under SFN.

In ISDB-T, it is possible to use DQPSK for symbol modulation like QPSK, 16QAM and 64QAM. DQPSK is more suitable for mobile reception than QPSK, 16QAM and 64QAM because it is very easy to compensate the channel characteristics. The depth of time interleave is variable within 0.5s. The longer

Table 11.4   Parameters of ISDB-T (6MHz Bandwidth).

ISDB-T mode	Mode 1	Mode 2	Mode 3
Number of OFDM segment	13		
Useful bandwidth	5.575 MHz	5.573 MHz	5.572 MHz
Carrier spacing	3.968 kHz	1.984 kHz	0.992 kHz
Total carriers	1405	2809	4992
Modulation	QPSK, 16QAM, 64QAM, DQPSK		
Number of symbols / frame	204		
Active symbol duration	252 μs	504 μs	1.008 ms
Guard interval duration	1/4, 1/8, 1/16, 1/32 of active symbol duration		
Inner code	Convolutional code (1/2, 2/3, 3/4, 5/6, 7/8)		
Outer code	Reed Solomon (204,188)		
Time interleave	0~0.5 s		
Useful bit rate	3.651 Mbps~23.234 Mbps		
Video Format	MPEG-2 Video		
Audio Format	MPEG-2 Audio (AAC)		

the time interval depth, the bit error rate characteristics under fading channels are improved. However, system latency will increase as the time interval depth is getting larger.

The parameters of guard interval length, inner code, code rate of inner code and outer code in ISDB-T are almost the same as those in DVB-T. Moreover, in ISDB-T, compensation of channel characteristics against QPSK, 16QAM and 64QAM modulation is performed using scattered pilot symbol similar to DVB-T.

The unique feature in ISDB-T is the use of BST-OFDM. In BST-OFDM, the frequency band (6 MHz) is divided into 13 segments that have equal bandwidth (about 430 kHz). These segments are combined to transmit programs and each group of segments can transmit different programs. Multi channel broadcasting and hierarchical broadcasting are easily realized by using BST-OFDM. The concept of BST-OFDM is shown in Figure 11.11.

In BST-OFDM, it is possible to transmit a HDTV program by combining 12 segments. Moreover, it is possible to transmit an audio program using the reminder one segment. As shown in Figure 11.11, the audio program segment is located in the center of the frequency band and it is possible to receive only an audio program using band pass filter. This form of reception is called partial reception and the receiver that performs partial reception is called a narrow

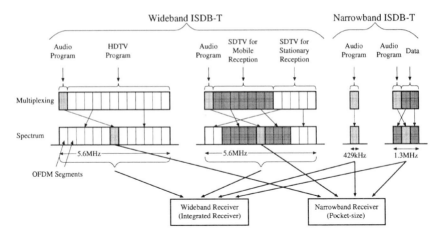

Figure 11.11   BST OFDM.

band ISDB-T receiver. The receiver that can process whole 13 segments is called a wide band ISDB-T receiver.

In ISDB-T, the form of broadcasting where maximum 3 segments are used for audio and data broadcasting is defined. This is called narrow band ISDB-T. Narrow band ISDB-T broadcasting can be received by both narrow band ISDB-T receivers and wide band ISDB-T receivers. The form of broadcasting where whole 13 segments are used is call wide band ISDB-T. Wide band ISDB-T broadcasting can be received only by a wide band receivers except for the case of partial reception.

In the example of multiplexing shown in the center of Figure 11.11, an audio program, an SDTV program for mobile reception and an SDTV programs for stationary reception are multiplexed. In the segments that transmit an SDTV program for mobile reception, DQPSK modulation is used for stable reception by sacrificing the quality of the program. In the segments that transmit an SDTV program for stationary reception, QAM is used for high quality reception using SFN. In this case, the same program can be transmitted in the different form of modulation and it is possible to select a suitable segments according to the form of reception. By using this technique, the broadcasting service that doesn't depend on the form of reception is realized and the chance of reception increases under many conditions. This is called hierarchical modulation. In BST-OFDM, the group of segments that transmits each program is called hierarchy. It is possible to use maximum three hierarchy in BST-OFDM. Symbol modulation scheme and code rate of inner code can be selected independently in each hierarchy. In DVB-T, as previously shown, multi channel

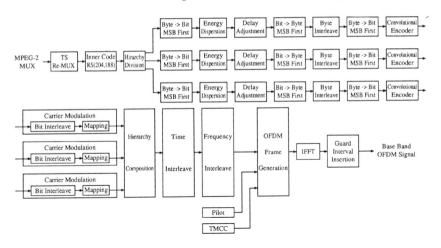

Figure 11.12 ISDB-T Wide Band Transmitter.

broadcasting or hierarchical broadcasting is performed by multiplexing before the transmitter.

The information of the hierarchical structure is transmitted by the TMCC (Transmission and Multiplexing Configuration Control) symbols inserted among the data symbols. The receiver configures its operation mode by demodulating TMCC symbols. TMCC is inserted in the same format independent of ISDB-T modes. TMCC is also used in DVB-T.

The structure of an ISDB-T transmitter is more complex than that of a DVB-T transmitter because of the adoption of BST-OFDM and time interleave. An example of a wide band ISDB-T receiver is shown in Figure 11.12.

Many field tests for ISDB-T are being performed in Japan toward the beginning of the services in 2003. The field experiments for mobile reception under MFN and SFN environments have already been performed and the validity of time interleaving is confirmed. Presently, there are 13 areas such as Tokyo, Osaka, etc. for pilot tests in Japan and various forms of broadcastings are under experiments. Since the structure of ISDB-T is very complex, development of low cost and high performance receivers is the subject which confronts us.

5   CONCLUSION

In this chapter, three principal digital terrestrial television broadcasting standards in the world are briefly described. Each standard is defined according to the requirements and circumstances of each country and it is difficult to say which is the best digital terrestrial television broadcasting standard at present.

As the result, these three standards will be scattered all over the world like analogue broadcasting. Therefore, interoperability between three standards is the subject in the future. Moreover, discussions about the contents of the service will be more and more necessary to increase the penetration rate.

The digitalization of broadcasting completes by the realization of digital terrestrial television broadcasting. After this, it is a big subject to investigate the next generation broadcasting that comes after digital broadcasting.

References

[1] ETSI "Digital Broadcasting Systems for Television, Sound and Data Services; Framing Structure, Channel Coding and Modulation for Digital Terrestrial Television," ETS 300 744 (1996).

[2] S. Moriyama "Present Situation of Terrestrial Digital Broadcasting in Europe and USA," the jounal of the ITE, **53**, 11, pp.1476–1478 (1999) (in Japanese).

[3] ATSC "ATSC Digital Television Standard," Doc. A/53 (1995).

[4] M. Kuehn, P. Chirist, C. Scarpa, S. Moriyama, M.itami " Digital Terrestrial TV Systems in the World," PIMRC'99 Panel Session #4 (1999).

[5] "MOTIVATE: Mobile Television & Innovative Receivers," T-Nova Deutche Telekom Innovationsgesellschaft mbH, Berkom.

[6] "Mobile Television an Innovative Receivers," http://b5www.berkom.de/MOTIVATE.

[7] M. Kawachi "The Recent Trend toward of Digital Broadcasting," the jounal of the ITE, **53**, 11, pp.1456–1459 (1999) (in Japanese).

Chapter 12

IMT-2000

Challenges of Wireless Millennium

Fumiyuki Adachi
Dept. of Electrical Communications, Graduate School of Engineering, Tohoku University
adachi@ecei.tohoku.ac.jp

Mamoru Sawahashi
Wireless Laboratories NTT Mobile Communications Networks, Inc.
sawahasi@mlab.yrp.nttdocomo.co.jp

Abstract   The 21st century will be a multimedia society, in which a combination of mobile communications and the Internet will play an important role. One good example is the success of the mobile Internet services called "i-mode services" provided by PDC (Japanese digital cellular standard) systems in Japan. Richer services will be provided by the third generation (3G) mobile communications systems, IMT-2000, which will be deployed around 2001-2002. Up to a 2-Mbps data transfer rate will be available and rich information, a mixture of text and images, will be transferred to mobile users with much better representation compared to present 2G systems. In this article, we see that providing Internet services will become of great importance. We look at how wireless access technologies are evolving and introduce IMT-2000 standardization activities targeting the global IMT-2000 or global 3G standard. Wideband DS-CDMA (W-CDMA) will be a major component of the global 3G standard. We introduce W-CDMA technology and present experimental results that show its effectiveness. Finally, we address advanced wireless techniques, i.e., interference cancellation and adaptive antenna array techniques that can enhance W-CDMA at a later date.

Keywords: IMT-2000, Mobile communications systems, multimedia, DS-CDMA

1   INTRODUCTION

Mobile communications systems have now become an important infrastructure of our society. Before the introduction of mobile communications systems, communications were only possible from/to fixed places, i.e., houses and offices. Mobile communications service started in December 1979 in Japan. For the first 10 years, its growth rate was very low. However, through the liberalization of mobile communications services in 1988 and terminal markets in 1994, the growth rate of mobile communications services accelerated. Similar rapid growth rates in mobile communications services are evident worldwide. An important factor that should not be overlooked is the increased utilization efficiency of portable phones (lighter weight and longer talk time, as seen in Fig. 12.1 [1], made possible by advanced LSI technology) and easier-to-use terminals. In Japan, the number of subscribers to cellular and Personal Handyphone System (PHS) services exceeded 55 million in February 2000; this number is equivalent to a penetration rate of 45 %. On the other hand, the number of fixed telephone users is continuously declining from its peak of 61 million in 1997. It was 58.5 million at the end of March 1999 and will be overtaken by the number of mobile communications users in March 2000. This clearly shows that people want to communicate with people, not with places. This is only possible through the aid of mobile communications technology. Mobile communications have enhanced our communications networks by providing an important capability, i.e., mobility.

Taking a looking at fixed communications networks, these networks are now not just for providing voice conversation and fax services. The rate at which the Internet communications services have proliferated throughout our society is striking. The Internet has proven itself to be a true driving force towards establishing a multimedia society in the 21st century. In line with the increasing popularity of Internet communications in fixed networks, mobile communications services have shifted their focus from solely voice conversation to electronic mailing and Internet access. One good example is the success of the mobile Internet access services called "i-mode services" provided by PDC (Japan digital cellular standard) systems in Japan [4]. This clearly indicates that the combination of mobile communications and the Internet will play an important role in this soon-to-arrive multimedia society. In the first years of the 21st century, a variety of new advanced services will be provided by the International Telecommunication Systems (IMT)- 2000 standardized in the International Telecommunication Union (ITU) [2].

In this paper, we show that mobile communications systems are now evolving from simply providing voice and fax communications services to providing Internet access services. We look at the trends of wireless access technologies, centering on the IMT-2000. We will introduce IMT- 2000 standardization ac-

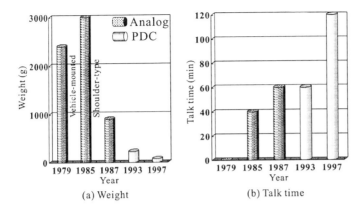

Figure 12.1   Advancement in weight and talk time of PDC mobile phones in Japanese markets.

tivities. W-CDMA will be a major component of the global IMT-2000 or the global 3G standard. We introduce W-CDMA technology and the supporting experimental results. We also address advanced wireless techniques, i.e., interference cancellation and adaptive antenna array, which can enhance W-CDMA at a later date.

2   BRIDGING THE EXPANSE OF MOBILE MULTIMEDIA: JAPANESE MARKET

2.1   21ST CENTURY INTERNET SOCIETY

In fixed networks, voice conversation was a long- time dominant service, but the introduction of Internet communications services has changed our society. Through the Internet, users can easily access WWW sites to retrieve various types of information including images, enjoy on-line shopping and trading services, and almost instantly exchange electronic mail messages instead of using traditional postal services. Information casting services represent another type of promising service. Internet communications services have been gaining popularity in our society with the aid of advancements in computer and data communication technologies. In Japan, the amount of Internet traffic is expected to surpass that of telephone traffic in 2001.

Figure 12.2 shows just how fast mobile, personal computers, and the Internet have grown in Japan [3]. It is evident that Internet services have spread throughout our society at a much faster speed than other services. Internet services took only 5 years to approach the 10% penetration (household) mark from the start of its commercial service (penetration rate was 11% by March 1999), while personal computers took 13 years to reach the same level. It

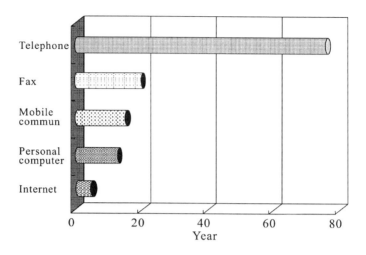

Figure 12.2   Time taken to arrive at 10% penetration (household) point.

is clear from the ever increasing popularity of Internet communications services in fixed networks that a combination of mobile communications, personal computers, and Internet services will drive our society to evolve into a mobile multimedia communications society in the 21st century.

2.2   MOBILE MULTIMEDIA COMMUNICATIONS

Our ultimate goal is to communicate *any information to anyone, at anytime, from anywhere*. The first steps toward bridging the expanse from today's society to the mobile multimedia communications society of the mid 21st century is seen in a new Internet access service called "i-mode services", which is provided over PDC-Packet Networks (Fig. 12.3) [4]. These services include e-mail, Web browsing, and various types of on-line services ranging from bank transactions to entertainment (Fig. 12.4). The i-mode terminals also allow conventional voice communications over PDC networks since PDC and PDC-P use the same TDMA air-interface. Since its introduction in Feb. 1999, i-mode's popularity has blossomed and close to 4 million users have subscribed to its services as of January 2000. Now, it seems that a mobile phone is not just for conversation, but is a communication tool that enables various types of electronic communications for private as well as business use. However, a slow data transfer rate (9.6 kbps in PDC-Packet air interface) and small displays in the portable phones allow access to only information written in the HyperText Markup Language (HTML) text format.

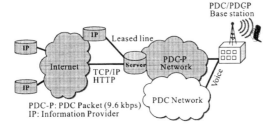

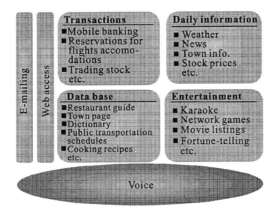

Figure 12.3 Multimedia-type "i-mode services" over PDC-Packet Networks.

Figure 12.4 Contents of "i-mode services".

3 EVOLUTION OF WIRELESS ACCESS TECHNOLOGY

3.1 OVERVIEW OF FIRST AND SECOND GENERATION SYSTEMS

Looking back through mobile communications history, we see the initial deployment of the first generation (1G) mobile communications systems, AMPS, TACS, NTT, etc., around 1980. These systems employed analog FM wireless access using frequency division multiple access (FDMA) with the channel spacing of around 25-30 kHz [5]. Then, the second generation (2G) systems, IS-54/136, GSM, and PDC, were deployed in the 1990's, all of which adopted time division multiple access (TDMA) with the channel spacing ranging from 25 to 200 kHz. Later, a new wireless access technique based on DS-CDMA appeared and IS-95 started its deployment [6]. Its channel spacing is much

Table 12.1   2G Systems.

	PDC (Mar. 1993)	GSM (1992)	TIA(USA) IS-136	TIA(USA) IS-95
Frequency (MHz)	800/1500	800/1900	800/1800/1900	
Wireless access	TDMA	TDMA	TDMA	CDMA
Carrier Spacing (kHz)	25	200	30	1256
No. of CH/carrier	6	8	3	Max. 64 (DL)
Speech Codec (kbps)	5.6	22.8	13	8 (variable rate)

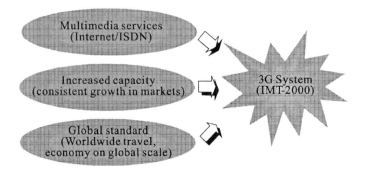

Figure 12.5   Preparing for a wireless multimedia society.

wider, i.e. 1250 kHz, compared to other 2G systems. In DS-CDMA, unlike in FDMA and TDMA, all users share the same frequency-band and time, but use different spreading code sequences to separate each user. 2G systems are listed in Table 12.1.

3.2   PREPARATION FOR A WIRELESS MULTIMEDIA SOCIETY

There are three strong reasons for developing IMT- 2000 systems: multimedia, higher capacity, and a global standard (Fig. 12.5).

All of the 1G and 2G systems are designed so that they can be optimized for basic services, i.e., voice, facsimile, and voice-band data. We have seen that major services provided by the 2G systems will shift from voice to multimedia communications over the Internet, as indicated by the "i-mode services". However, the data transfer rate is around 9.6 kbps, which is far too slow for retrieving content-rich information comprising text and images. Users will demand much higher transfer rates and much better representation on the phones. In multimedia communications, supporting multi-rate and variable-rate communications is a paramount requirement and the data-rate range will be significantly wide, e.g., as low as 8 kbps to a couple of megabits per second. Furthermore, in order to cope with the still-continuing rapid growth of mobile communications, the issue of capacity must also be addressed. Finally, establishing a global standard is becoming an increasingly important issue in the 21st century, when more and more people will travel around the world for businesses and leisure (2G system standards are, more or less, regional standards).

3.3   IMT-2000

IMT-2000 systems are expected to be deployed worldwide starting around 2001-2002. They will be operated in all radio propagation environments from outdoors to indoors, urban and suburban areas, and hilly and mountainous areas. Minimum requirements in terms of data transfer rates and quality for different environments are summarized below.

- Indoor: 2.048 Mbps and bit error rate (BER) = 10^{-6}

- Pedestrian: 384 kbps and BER = 10^{-6}

- Vehicular: 144 kbps and BER = 10^{-6}

Data transfer rates of up to 2 Mbps and the same quality as fixed networks are the targets. For the transmission of image information of 1 Mbyte, 14 min. is necessary at a 9.6 kbps user rate, but the transmission time will be significantly shortened (to 4 sec.) with a 2 Mbps transfer rate. However, it is a difficult challenge to realize this high rate and high quality transmission in harsh mobile communication channels. Advanced wireless techniques must be developed.

Figure 12.6 summarizes the evolution path from the 1G to 3G systems. Interestingly, every decade, new technology has emerged that enhances the communications capability.

3.4   SPECTRUM ISSUE

The available radio spectrum is a key factor for success of the IMT-2000 systems since the bandwidth limits the available user rates and the frequency band influences the terminal costs as well as talk and standby time of mobile phones. At the World Administrative Radio Conference (WARC) 1992, a 230-MHz spectrum in the 2-GHz band was identified for the IMT-2000 (Fig. 12.7). Basically, Europe and Japan will follow the recommendations, while a significant part of the WARC spectrum in the lower bands has been allocated to the

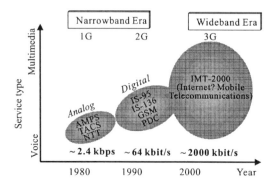

Figure 12.6   Evolution path.

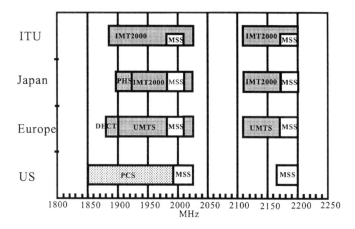

Figure 12.7   Frequency allocation by ITU and present frequency plans of Japan, Europe, and USA.

personal communications services (PCS) systems in the USA. Due to recent rapid growth in mobile communications and the demand for a broad range of multimedia communications services, additional IMT-2000 spectra after the initial deployment are requested. This will be discussed at World Radio communication Conference (WRC) 2000.

4   IMT-2000 STANDARDIZATION

4.1   NEED FOR NEW WIDEBAND WIRELESS ACCESS TECHNIQUE

Study on the IMT-2000 (former FPLMTS) was initiated by the ITU in 1986. The system concept and requirements were set by the ITU and standardization activities have been intensified worldwide as we enter the new millennium [7].

As already mentioned, major services provided by the IMT-2000 system will be multimedia communications over the Internet such as the "i-mode services", but at much higher transfer rates and with a better representation compared to the 2G systems. These services will be wideband multimedia, Internet access, imaging and videoconferencing, as well as basic services (voice, fax, and voice-band data). To realize such a system, a new wideband wireless access technique incorporating as many recent technological achievements as possible is necessary.

Since the appearance of the wireless DS-CDMA technique, a heated debate has continued regarding which access technique, TDMA or DS-CDMA, provides a larger link capacity. However, it is quite a difficult task to conclude this debate since the link capacities offered by these techniques are different under different assumptions. Nevertheless, wireless DS-CDMA has numerous advantages over FDMA and TDMA, e.g., single frequency reuse, soft and softer handoff, and Rake combining. Soft handoff improves the transmission quality at places near cell boundaries. Widening the signal bandwidth creates a serious problem of intersymbol interference (ISI) due to frequency selective multipath fading; however, Rake combining exploits frequency selective multipath fading and can improve significantly the transmission performance. The intensive research on DS-CDMA worldwide has proven that widening the spreading bandwidth is the best way to fulfill the requirements for the IMT-2000.

In addition to the above, one important advantage of DS-CDMA is that multi-rate variable-rate transmission, required for providing multimedia services, can be easily established by changing the spreading factor or code multiplexing while keeping the spreading bandwidth the same.

4.2   STANDARDIZATION ACTIVITIES

In Japan, the Association of Radio Industries and Businesses (ARIB) started in 1995 a selection process for a wireless access technique and choose W-CDMA in January 1997. Since then, the ARIB has been actively promoting W-CDMA worldwide for its acceptance as a global standard [8]. W-CDMA consists of both the frequency division duplex (FDD) and time division duplex (TDD) components. The TDD component was designed based on the

Table 12.2   RTT Proposals to ITU.

DECT	Digital Enhanced Cordless Telecommunications	ETSI Project DECT
UWC-136	Universal Wireless Communications	USA TIA TR 45.3
WIMS	Wireless Multimedia and Messaging	USA TIA TR 46.1
W-CDMA	Services Wideband CDMA	
TD-SCDMA	Time Division Syncronous CDMA	China Academy of Telecom. Technol.
W-CDMA	Wideband DS-CDMA	Japan ARIB
CDMA II	Asynchronous DS-CDMA	S. Korea TTA
UTRA	UMTS Terrestrial Radio Access	ETSI SMG2
NA: W-CDMA	North American Wideband DS-CDMA	USA T1P1-ATIS
cdma2000	Multiband synchronous DS-CDMA	S.Korea TTA

same concept as the FDD component so that as much commonality between the FDD and TDD terminals can be obtained. Taking advantage of the fact that the forward link fading and reverse link fading are highly correlated, the TDD component can adopt open-loop fast transmit power control and transmit/receive antenna diversity at a cell site (transmit antenna selection can be based on channel state information on the receive antennas) [9].

In Europe, the European Telecommunications Standards Institute (ETSI) arrived at an historic decision on UMTS Terrestrial Radio Access (UTRA) in January 1998 which was to adopt W-CDMA for the FDD bands and adopt a hybrid solution of TDMA and CDMA called TD-CDMA for TDD [10]. In TD-CDMA, users are assigned a code and time slot. Multiple codes and slots can be assigned to a user for higher rate services. A sophisticated joint detection algorithm is adopted on the reverse link to allow multiple users within one time slot while relaxing the power control accuracy [11]. Meanwhile in the United States, TIA prepared several proposals including cdma2000 as a DS-CDMA evolution from IS-95 (also among the proposals was UWC-136 as a TDMA evolution from IS-136).

4.3   RTT PROPOSALS

ITU-R TG8/1 called for radio transmission technology proposals by June 1998 and for system evaluation reports by September 1998 [12]. A total of ten proposals for a terrestrial access technique were submitted to the ITU. Eight of the ten proposals were based on DS-CDMA as shown in Table 12.2. Since then, W-CDMA has been recognized as the strongest candidate for the IMT-2000 air interface and the development of systems based on W-CDMA has

Table 12.3   Japan's W-CDMA Proposal to ITU (June 1998).

Bandwidth (MHz)	1.25/5/10/20
Chip rate (Mc/s)	1.024/4.096/8.192/16.384
Duplex scheme	FDD/TDD
Spreading Code	OVSF code + Scramble code
Modulation (FL/RL)	QPSK/QPSK
Data modulation (FL/RL)	FDD: QPSK/BPSK TDD: QPSK/HQPSK
Data detection (FL)	Coherent detection based on FDD/TDD: TCH dedicated time multiplexed PL
Data detection (RL)	Coherent detection based on FDD: I/Q multiplexed pilot TDD: Time-multiplexed pilot
Frame length	10 ms
Multi-rate	Variable SF and multicode
Channel coding	Convolutional codes (R=1/3 and 1/2, K=9) Turbo codes (R=1/3)
Inter-cell timing	FDD: Asynchronous (Synch. possible) TDD: Synchronous

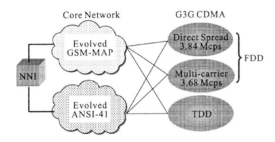

Figure 12.8   Harmonized solution to global 3G.

accelerated throughout the world. Table 12.3 shows the W-CDMA proposal from Japan (June 1998) to the ITU.

4.4   HARMONIZATION EFFORTS TOWARD A GLOBAL STANDARD

Intensive harmonization studies between W-CDMA and cdma2000 were conducted to establish a global 3G standard, which will lend large-scale economic advantages to consumers, network operators, and manufacturers. A global 3G standard was agreed upon in 1999 [13]. In the harmonized global 3G standard, there will be three DS-CDMA operation modes: FDD single-carrier, FDD multi-carrier, and TDD (Fig. 12.8). The FDD single-carrier mode will

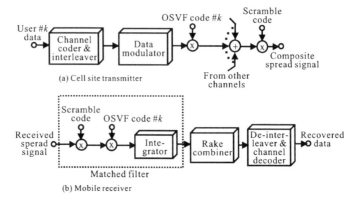

Figure 12.9   Forward link transmission and reception process.

be based on the W-CDMA proposal but with a chip rate of 3.84 Mcps, while the FDD multi-carrier mode will be based on the cdma2000 proposal. There are two fully globally-established core networks presently used for the 2G systems: GSM-MAP and ANSI-41. The former is used by GSM systems and the latter is used by AMPS and IS-95 systems. Both core networks will evolve into 3G systems and the above mentioned air-interfaces must connect to both GSM- MAP and ANSI-41 core networks.

5   W-CDMA WIRELESS ACCESS

The important concepts of the original W-CDMA proposal from Japan are the introduction of inter-cell *asynchronous* operation, the dedicated pilot channel, and multi-rate transmission. Below, the unique technical features developed thus far for this original W-CDMA are summarized (the concept behind the harmonized solution of FDD is quite similar to the original proposal, but modified through the harmonization process).

- Fast cell search algorithm for inter-cell *asynchronous* operation [14]

- Orthogonal variable spreading factor (OVSF) codes on forward links [15]

- Coherent Rake receiver [16]

- Signal-to-interference ratio (SIR)-based fast transmit power control (TPC) scheme on both reverse and forward links [17]

- Variable rate transmission with blind rate detection [18]

- Multi-stage interleaver for Turbo coding [19]

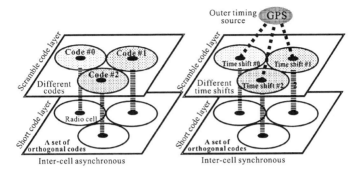

Figure 12.10 Two-layered code structure.

Figure 12.9 illustrates the forward link transmission process and reception process. User data to be transmitted are first channel coded and data modulated, then spread to a wider bandwidth, e.g., 5 MHz, by applying the OVSF codes. The spread signals of all users are summed and scrambled by the cell-site unique scramble code to be transmitted from the antenna after power-amplification. Since multipath propagation channels comprise several paths with different time delays, several copies of the transmitted spread signal are received at a mobile station. At the mobile receiver, each copy of the transmitted spread signal is matched filtered using the regenerated OVSF codes to obtain a copy of the transmitted data modulated signal. A Rake combiner then combines all of the copies of the data modulated signal to obtain a soft decision data sequence for successive channel decoding to recover the transmitted data sequence.

The forward and reverse links have a frame structure; each frame the length of which is 10 ms is divided into 16 slots of 0.625 ms. One slot corresponds to one power-control period.

5.1 INTER-CELL ASYNCHRONOUS OPERATION

The inter-cell asynchronous operation allows easier system deployment from outdoors to indoors because no external timing source such as the Global Positioning System (GPS) as used in the inter-cell synchronous system, i.e., IS-95 is required. Unlike inter-cell synchronous systems, a two-layered code structure (Fig. 12.10) is adopted and each different cell site has a unique scramble code sequence assigned to its forward link. In general, the inter-cell asynchronous operation increases cell search time at a mobile station (time spent before finding and synchronizing to the best cell site to access). This problem is overcome by a 3-step cell search algorithm [14]. The scramble codes used in

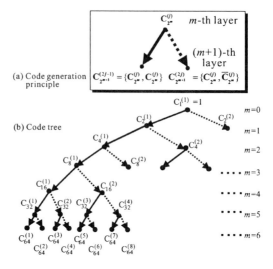

Figure 12.11   OVSF code tree.

the system are grouped into subgroups and each subgroup is represented by the group identification (GI) code. From each cell site, a slot timing code (common to all cells) and GI code representing its scramble code are periodically transmitted in parallel. The 3-step algorithm comprises:

Step 1: slot timing detection using matched filtering

Step 2: search for GI code and frame timing

Step 3: search for scramble code within GI code group

According to our measurements, the 3-step algorithm accomplishes the cell search within 960 ms at a 90 % probability for 512 scramble codes [20].

The harmonized global 3G (G3G) FDD solution also supports inter-cell synchronous operation as well as inter-cell asynchronous operation in order to deploy the G3G FDD systems more flexibly in different environments.

5.2   OVSF CODES

On the forward link, all user signals are time synchronous, orthogonal spreading can be used to mitigate the multiple access interference (MAI). However, as the frequency selectivity of the propagation channel becomes stronger (or the number of resolvable paths of the propagation channel increases), the orthogonality among different users tends to diminish because of increasing inter-path interference. Nevertheless, orthogonal spreading always gives a larger link capacity than random spreading.

Data transmission at the symbol rate that equals the chip rate/2^m is done by using a single code with the spreading factor (SF) of 2^m, where m is a positive integer. However, a question arises as to how to establish orthogonality among users transmitting at different data rates or symbol rates while keeping the spreading code chip rate the same (or keeping the spreading bandwidth the same). This can be achieved by using the OVSF codes. These codes can be generated recursively based on a modified Hadamard transformation, resulting in a tree-structured code set shown in Fig. 12.11. Starting from, a set of 2^m orthogonal spreading codes is generated at the m-th layer ($m = 1, 2, 3, ...$) from the top. The generated spreading codes are Hadamard-Walsh codes. However, code generation can start from any m-th layer using a set of 2^m orthogonal codes (e.g., a set of orthogonal Gold codes if it exists) other than Hadamard-Walsh codes. The code length of the m-th layer is 2^m chips and can be used for transmitting symbols at the rate of 2^m times lower than the chip rate.

5.3 COHERENT RAKE RECEIVER

Since the DS-CDMA links are interference-limited, the link capacity is almost inversely proportional to the required signal energy per information bit-to-interference plus background noise power spectrum density ratio E_b/I_0. Coherent detection can reduce the required E_b/I_0. However, in general, coherent detection, which requires accurate channel estimation is quite difficult to achieve in severe channel conditions, i.e., fast fading and low received signal powers. One efficient way to achieve this is to use the dedicated pilot channel to transmit the information-nonbearing symbols for channel estimation. The transmitting pilot channel seems to incur energy loss, but improves significantly the overall transmission performance in fading environments, compared to non-coherent detection. Furthermore, it is easier to adopt advanced techniques, i.e., interference cancellation and adaptive antenna array techniques, at a later date; they are addressed later. For the time-multiplexed pilot channel case, a $2K$-tap weighted multi-slot averaging (WMSA) channel estimation filter can be applied [16]. The coherent Rake receiver structure using WMSA channel estimation filter is illustrated in Fig. 12.12.

5.4 FAST TPC BASED ON SIR MEASUREMENT

All users are time-*asynchronous* on the reverse link since all users are transmitting from different locations. In this situation, severe MAI is produced due to distance-dependent path loss (this is well known as the near/far problem), random path loss or shadowing, and multipath fading. In order to minimize the MAI, fast TPC is indispensable in that it controls the mobile transmit powers so that all user signals are received at the same power at the cell site. Since the DS-CDMA channels are interference-limited, the fast TPC should be based on

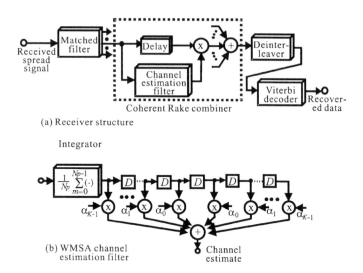

(a) Receiver structure

(b) WMSA channel estimation filter   Channel estimate

Figure 12.12   Coherent Rake receiver.

the SIR measurement. The structure of fast TPC is shown in Fig. 12.13. The mobile transmit power is raised or lowered by a certain amount if the measured SIR at the cell site receiver is less or greater than the target SIR. The outer loop control updates the target SIR so that the received frame error rate (FER) is maintained at the required value. Since the fast TPC is performed user by user, the target SIR for each user is not necessarily the same. It depends on the data rate and the required FER. SIR-based fast TPC can always minimize the mobile transmit powers according to variations in the traffic load. This type of fast TPC can be applied to the forward link, i.e., power-control of the cell site transmit powers [21]. It should be pointed out that the combined use of fast TPC and coherent Rake yields the greatest improvement in the transmission performance in frequency selective fading environments. Thus, interference to other users in other cells can be reduced, thereby increasing the link capacity.

Present fast TPC has a constant power up/down step size. However, introducing adaptability to the power step size may further reduce the TPC error under some propagation conditions [22].

5.5   VARIABLE RATE TRANSMISSION WITH BLIND RATE DETECTION

The data rate may change frame-by-frame during the communication. Since DS-CDMA links are interference-limited, variable rate transmission can reduce the average interference power, thus contributing to increasing the link

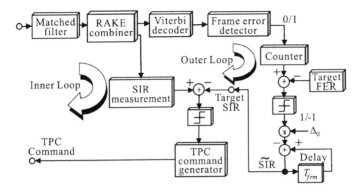

Figure 12.13   SIR-based fast TPC.

capacity. In the harmonized solution of FDD, the rate information is transmitted frame-by-frame (blind rate detection is used only for voice transmission on the forward link). If the set of possible rates is known to the receiver, the transmitted variable rate data can be recovered without transmitting the rate information (blind rate detection). This will be introduced below.

One realization of variable rate transmission is to fill partially each slot (in a frame) when the data rate is below the maximum while keeping the symbol rate the same (resulting in discontinuous transmission). In this case, blind rate detection can be incorporated into the structure of Viterbi decoding of coded frame data, and cyclic redundancy check (CRC) decoding is used to determine whether recovered variable rate data is correct. Another realization of variable rate transmission is to vary, according to the data rate, both the transmit power and the spreading factor while keeping the chip rate the same; discontinuous transmission can be avoided. The data symbol sequence is further modulated (data-independent) after data modulation but before spreading so that double-modulated symbol sequences with possible rates are all orthogonal to each other; at the receiver, noncoherent orthogonal demodulation is applied for rate detection. Therefore, no rate information needs to be transmitted.

5.6   MULTI-STAGE INTERLEAVER FOR TURBO CODING

Turbo coding [23] is characterized by:

- Parallel concatenation of two or more recursive systematic codes (RSCs)

- Interleaver with RSC

- Iterative decoding algorithm

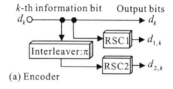

(a) Encoder

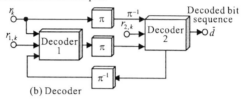

(b) Decoder

Figure 12.14   Rate-1/3 Turbo encoder and decoder.

Turbo coding is particularly useful for high-speed, high quality data transmission in W-CDMA [24]. A simplified structure of a turbo encoder/decoder is illustrated in Fig. 12.14. In the iterative decoding of turbo codes, the current soft-in/soft-out decoder computes extrinsic information that plays an important role in the turbo decoding and feeds them back to the next soft-in/soft-out decoder to be updated. In the turbo decoding process, the random property of the interleave pattern and large separation of the nearest input bits are important. To achieve this, a multi-stage block interleaver (MIL) [19] that interleaves row and column data recursively is applied.

Figure 12.15 plots the simulated BER performance of 64-kbps data transmission over W-CDMA reverse link (1/3-rate turbo coding, SF of 32, and chip rate of 4.096 Mcps). In the simulation, the turbo coding interleaver size was 80 ms (5773bits). The ITU-R Vehicular-B fading channel model having 6 Rayleigh faded paths with $f_D = 80$ Hz, antenna diversity, 4-finger Rake, and fast TPC were assumed. The figure shows that the MIL interleaver provided a larger coding gain than did the random interleaver. For comparison, the simulation results for rate-1/3 convolutional coding (cc) with constraint length of 9 bits and concatenated coding (cc+RS) comprising a Reed-Solomon outer code followed by a convolutional inner code are also plotted. A larger coding gain is obtained by Turbo coding.

6   EXPERIMENTAL EVALUATION OF W-CDMA

Results of field experiments on a 32-kbps data transmission over a W-CDMA reverse link channel are presented below. The carrier frequencies of the reverse (mobile-to-base) and forward (base-to- mobile) links were 1.9905/2.175

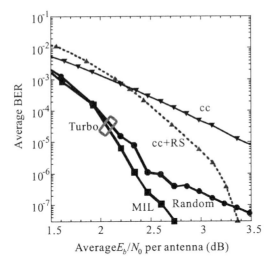

Figure 12.15 BER performance with Turbo coding.

Measurement van

CDMA transceiver

Figure 12.16 Measurement van equipped with W-CDMA transceiver.

GHz, respectively. The chip rate and the symbol rate were 4.096 Mcps and 64 ksps (SF of 64), respectively. This experiment applied a rate-1/3 convolutional coding and soft decision Viterbi decoding. Figure 12.16 are photos of a measurement van equipped with a mobile transceiver. The measurement van was

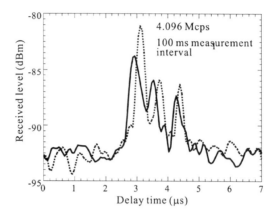

Figure 12.17   Measured power delay profile.

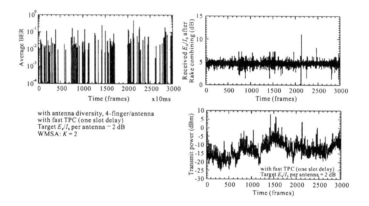

Figure 12.18   Example of time variations in received BER, E_b/I_0 and transmit power.

driven along the measurement courses, the distance of which ranged from 0.5 to 1 km from the base station, at an average speed of approximately 30 km/h (the maximum Doppler frequency of fading was approximately 55 Hz). The measured power delay profiles of the multipath channel along the test courses are plotted in Fig. 12.17. Two to three distinct propagation paths with unequal powers were observed. The received spread signals that were propagated along these propagation paths were despread and coherently combined by a Rake combiner for successive channel decoding and BER measurement.

Examples of the time variations of the measured average BER, the received signal energy per information bit-to-interference plus background noise power

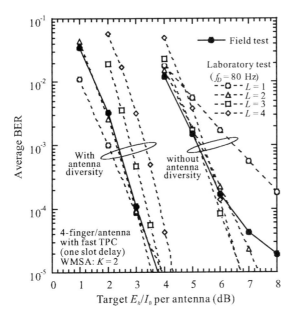

Figure 12.19   BER performance with fast TPC and Rake.

spectrum density ratio (E_b/I_0) after Rake combining, and mobile transmit power, each measured every 10 ms (1 data frame) interval, are shown in Fig. 12.18. The target E_b/I_0 per antenna was set to 2 dB (outer loop control was not used). The mobile transmit power varied with the range of about 35 dB, so that the received E_b/I_0 after Rake combining was satisfactorily brought to almost a constant value. The short-term average (10 ms) BER was below 10^{-3} in most regions of the measurement course. The measured average BER performance of the reverse link is plotted in Fig. 12.19. With antenna diversity reception, the average BER of 10^{-3} can be achieved at the average E_b/I_0 of approximately 3 dB per antenna, resulting in increased link capacity and coverage.

7   W-CDMA ENHANCEMENT

In any cellular DS-CDMA system, since all users in different cells use the same carrier frequency, the links are not only power-limited but also interference-limited. To increase the link capacity in this situation, we first identify the W-CDMA enhancing techniques using a simple capacity equation, and then address the most promising techniques.

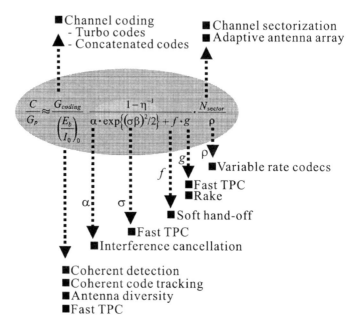

Figure 12.20   Enhancing techniques.

7.1   IDENTIFYING ENHANCING TECHNIQUES

The techniques that increase the link capacity are identified by using the following reverse (mobile to base) link capacity, C, normalized by processing gain G_p [25].

$$\frac{C}{G_p} \approx \frac{G_{coding}}{\left(\frac{E_b}{I_0}\right)_0} \cdot \frac{1 - \eta^{-1}}{\alpha \exp\{(\sigma\beta)^2/2\} + f \cdot g} \cdot \frac{N_{sector}}{\rho},$$

where $(E_b/I_0)_0$ is the required E_b/I_0 without channel coding and with perfect fast TPC, G_{coding} is the channel coding gain, α represents the interference per user-to-desired user power ratio, σ the fast TPC error, $\beta = 10/\ln 10$, f the outer cell interference-to-own cell interference power ratio with no multipath fading (only distance dependent and shadowing are considered), g the power rise factor due to fast TPC when multipath fading is considered, ρ the average-to-peak transmission rate (variable rate case), N_{sector} the sectorization factor, and finally η the allowable power increase factor from the background noise. Many techniques are identified from the above equation for increasing the link capacity. They are summarized in Fig. 12.20. Techniques may fall into two technical areas: interference reduction techniques and required E_b/I_0 reduc-

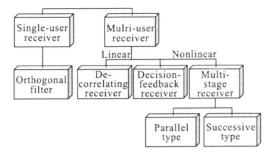

Figure 12.21   Family of interference cancellation techniques.

tion techniques. The former includes interference cancellation and the adaptive antenna array. Advanced coherent Rake combining, advanced channel coding, and advanced fast TPC may fall into the latter area. However, the most powerful and interesting advanced techniques are interference cancellation and the adaptive antenna array, which can directly reduce the interference power.

The effectiveness of these techniques is explained below. If the interference within the own cell is perfectly cancelled, the link capacity can increase by $(\alpha + f)/f$. In the case of $\alpha = 1$, $\sigma = 0$, and $f = 0.71$, the capacity can be increased by 2.4 fold. Cell sectorization uses sector antennas instead of an omni antenna. Since the total interference power can be reduced by a factor of N_{sector}, the number of sectors/cell, the capacity can be increased N_{sector} fold (because of the beam side-lobe spill-over to adjacent sectors, the actual capacity increase is smaller than the number of sectors/cell). If a beam pattern is generated for each user, this technique is called an adaptive antenna array.

7.2   INTERFERENCE CANCELLATION (IC)

IC techniques are classified in Fig. 12.21. Basically, there are two types: the single-user detection type and the multi-user detection type [26]. The orthogonal matched filter (MF) receiver is one example of the single-user detection type IC, which controls the MF tap coefficients so that the desired signal component is made orthogonal against other users' spread signals, and is simpler to implement than the latter. However, W-CDMA uses long random spreading code sequences on the forward link for scrambling and on the reverse link. Thus, the time-varying nature of a spreading code sequence, when observed over each one-symbol period, excludes adoption of the single-user type.

Multi-user IC receivers can be classified into linear and non-linear types. The decorrelating receiver is the linear type multi-user IC receiver. However, since the computation complexity grows exponentially with the increased number of users and the length of the spreading code sequence, it is considered to

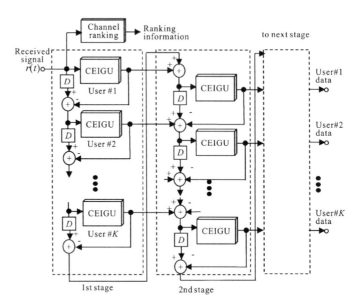

Figure 12.22   Multistage IC receiver structure.

be impractical. On the other hand, the decision-feedback IC receiver and the multistage IC receiver [27] are non-linear type receivers. The former is the combination of the decision-feed back equalizer and the successive IC. The multistage IC receiver with non-linear replica generation is considered to be most attractive for W-CDMA mobile radio applications. This is because MAI is subtracted successively from each canceling stage in the order of decreasing signal power from the strongest user, and consequently, more accurate interference replica generation and subtraction is possible than parallel cancellation.

In multi-stage IC, accurate channel estimation is necessary to generate the interference replica of each user. This is again accomplished by taking advantage of the dedicated pilot channel introduced for coherent Rake combining. The resulting multi-stage IC receiver is called the coherent multistage IC (COMSIC hereafter) receiver [28, 29]. The simplified structure of the COMSIC receiver is illustrated in Fig. 12.22. The channel estimation-and-interference replica generation unit (CEIGU) performs matched filtering (de-spreading), channel estimation, tentative symbol decision, and interference replica generation for each user. Interference replica generation and subtraction are done successively in decreasing order of the received signal powers and hence, the accuracy of channel estimation can improve for lower ranked users. Furthermore, the channel estimate is updated at successive stages for

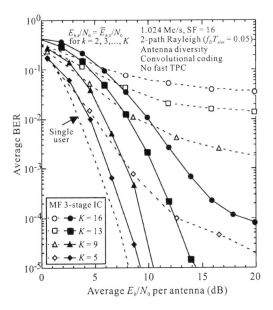

Figure 12.23   Measured BER performance with 3-stage IC.

each user so that the accuracy of the channel estimation improves with the stage number.

BER performance improvement is saturated at the 3rd stage. The measured average BER performance of an implemented 3-stage COMSIC receiver (1.024 Mcps, SF = 16) is plotted in Fig. 12.23. As the average E_b/N_0 increases, the average BER with the IC receiver monotonically falls, while that with the MF receiver approaches an error floor. Even when K = 16 users exist (which equals the value of SF), an average BER below 10^{-3} can be achieved at the average E_b/N_0 per antenna of approximately 14 dB.

7.3   ADAPTIVE ANTENNA ARRAY

The conceptual structure of the adaptive antenna array receiver is illustrated in Fig. 12.24. It directs beam nulls toward interference sources to maximize the signal-to-interference plus background noise (SIR) of each user [30, 31, 32, 33, 34]. In the case of voice-only services, since the required SIR is the same for all users, a large number of antenna elements are required and thus, the application of the adaptive antenna array receiver is considered to be rather impractical. This is the reason why the adaptive antenna array was mainly considered for TDMA mobile radio applications. However, it is also useful in the multimedia DS-CDMA mobile radio because different users are transmitting at

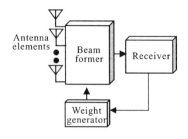

Figure 12.24   Adaptive antenna array.

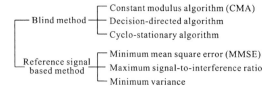

Figure 12.25   Optimum weight-generation criteria of beam former.

different data rates and the reduction of the interference power from high rate users directly results in the capacity increase. Even the use of a small number of antenna elements may significantly increase the capacity. Since the adaptive antenna array requires very complicated signal processing, it was considered impractical in the past. However, due to recent advancements in DSP technology, the adaptive antenna array can be considered practical in a few years.

Figure 12.25 lists the criteria for generating optimum antenna beam former weights [30]. There are basically two types: blind type and reference signal based type. The constant modulus algorithm (CMA) is a well-known blind algorithm that requires no reference signal; however, it has a drawback in that any signal with the maximum received signal power can become the desired signal. The reference signal based algorithm requires an accurate reference signal. In W-CDMA, the reference can be extracted using the pilot channel. There are three algorithms: minimum mean square error (MMSE), maximum SIR, and minimum variance. The generated antenna weights using these different criteria are all given by the Wiener solution.

In the DS-CDMA, adaptive beam forming (space domain processing) and Rake combining (time domain processing) must be considered. The optimal solution is to combine adaptive beam forming and Rake combining functions; however, this requires a high degree of complexity. A pragmatic solution is to completely separate the adaptive beam forming and Rake combining functions [34]. The structure of our developed coherent adaptive antenna array

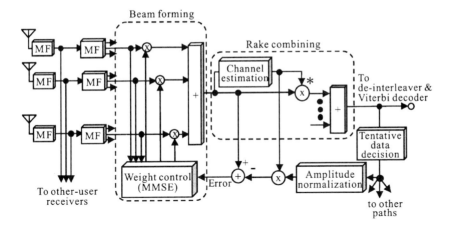

Figure 12.26 Coherent adaptive antenna array diversity receiver.

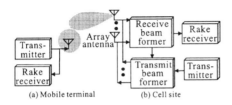

Figure 12.27 Adaptive antenna array transceiver at cell site.

diversity is shown in Fig. 12.26. The antenna beams need to track only slow changes in the arrival angles and the average powers of the desired and interfering signals so that the average received SIR corresponding to each user is maximized. It is the Rake combiner's task to track the fast changes in the received signals due to multipath fading and coherently combine the resolved desired signal components to maximize the instantaneous SIR.

Since receive antenna beam forming does not involve instantaneous information regarding the multipath channel parameters, the receive antenna weights are carrier frequency-independent. This suggests that the transmit antenna beam forming can be based on the receive antenna weights (however, appropriate calibration due to RF circuit amplitude/phase differences among different antenna branches is necessary). This makes it possible to adopt the adaptive antenna array only at the cell site as shown in Fig. 12.27. On the reverse link, both the beam former and Rake combiner are equipped at the base sta-

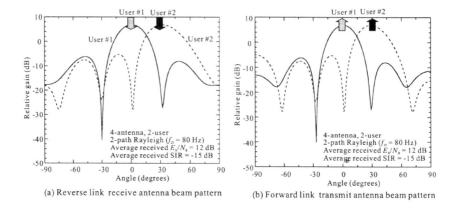

Figure 12.28   Receive and transmit antenna beam patterns.

tion, while on the forward link the beam former and the Rake combiner are equipped at the base station and the mobile station, respectively.

The measured receive (reverse link) and transmit (forward link) antenna beam patterns of the implemented antenna array receiver are plotted in Fig. 12.28 for W-CDMA (4.096 Mcps, SF = 64). Beam nulls can be directed toward the interfering users. Average BER measurements confirmed that adaptive antenna array reception can work satisfactorily in a multipath fading environment and yields superior performance to antenna diversity reception, and adaptive antenna array transmission improves the forward link performance.

8   CONCLUSION

Mobile radio communication systems are about to evolve into mobile multimedia communications systems that can flexibly offer various types of Internet services to mobile users. The third generation mobile communications system, IMT- 2000, is expected to play an important role in this soon-to-arrive mobile multimedia society. IMT-2000 systems will be based on wideband CDMA technology. Up to a 2 Mbps data transfer rate will be available and content-rich information comprising text and images will be transferred to mobile users at a much faster rate with a much better representation compared to present 2G mobile terminals with small, mono color screens. This is a significant advancement.

Almost every decade, a new generation system has appeared. Based on this, we predict that the fourth generation system will emerge around 2010. The fourth generation system should be a broadband packet wireless system optimized to the next generation Internet protocol, probably with maximum rates of more than 2 Mbps in a vehicular environment and 10-20 Mbps in stationary-to-pedestrian environments [35]. However, this is quite a difficult technical challenge since wireless channels become quite adverse due to dense multi-path environments. The frequency bands for the fourth generation systems will most likely lie above 5 GHz. We must remember that propagation loss is in proportion to 2.6^{th} power to the carrier frequency, i.e., $\propto f_c^{2.6}$. Therefore, the radio links are not only interference-limited but also severely power-limited. The adaptive antenna array plays a key role in abating this power problem. MAI (or collision) also limits the packet throughput, so packet interference rejection or employing a packet IC receiver is also an important technique. Due to the micro/pico-cell structure, it is quite difficult for the fourth generation system to provide nationwide coverage. Only high-traffic areas may be covered. Close cooperation with other systems, e.g., IMT-2000 system is thus, necessary. This requires the so-called software radio technology to enable a single mobile terminal to access both third and fourth generation systems. We would like to emphasize that very difficult but interesting technical challenges still wait for us in the coming years.

References

[1] K. Honma et al., "Mobile terminal technologies," Proc. IEICE, vol. 82, pp. 138-144, Feb. 1999.

[2] http://www.itu.int/imt/

[3] Communications White Paper, MPT, Japan, 1999.

[4] Special issue, i-mode service, NTT DoCoMo Technical Journal, vol. 7, July 1999.

[5] Special issue, Wireless Personal Communications, IEEE Commun. Mag., vol. 33, Jan. 1995.

[6] K.S. Gilhousen, I.M. Jacobs, R. Padovani, A.J. Viterbi, L.A. Weaver, and C.E. Wheatley III, "On the capacity of a cellular CDMA system," IEEE Trans. Veh. Technol., vol. VT-40, pp. 305-312, May 1991.

[7] Special issue, IMT-2000: Standards efforts of the ITU, IEEE Personal Commun., vol. 4, Aug. 1997.

[8] Special issue, Next generation mobile communications, Proc. IEICE (in Japanese), vol. 82, pp. 101-160, Feb. 1999.

[9] K. Miya, et al, "Wideband CDMA systems in TDD-mode operation for IMT-2000," IEICE Trans. Commun., vol. E81-B, pp. 1317-1326, July 1998.

[10] Special issue, Wideband CDMA, IEEE Commun. Mag., vol. 36, Sept. 1998.

[11] A. Klein, G.H. Kaleh, and P.W. Baier, "Zero forcing and minimum mean square error equalization for multiuser detection in code division multiple access channels," IEEE Trans. Veh. Technol., vol. 45, pp. 276-287, May 1996.

[12] ITU-R, Request for submission of candidate transmission technologies (RTTs) for IMT- 2000/FPLMTS radio interface, circular letter 8/LCCE/47, 1997.

[13] P. Chaudhury, W. Mohr, and S. Onoe, "The 3GPP proposal for IMT-2000," IEEE Commun. Mag., vol. 37, pp. 72-81, Dec. 1999.

[14] K. Higuchi, M. Sawahashi, and F. Adachi, "Fast cell search algorithm in DS-CDMA mobile radio using long spreading codes," IEICE Trans. Commun., vol. E81-B, July 1998.

[15] F. Adachi, M. Sawahashi, and K. Okawa, "Tree-structured generation of orthogonal spreading codes with different lengths for forward link of DS-CDMA mobile radio," IEE Electron. Lett., vol. 33, pp. 27-28, Jan. 1997.

[16] H. Andoh, M. Sawahashi, and F. Adachi, "Channel estimation using time-multiplexed pilot symbols for coherent Rake combining for DS-CDMA mobile radio," IEICE Trans. Commun., vol. E81-B, July 1998.

[17] S. Seo, T. Dohi, and F. Adachi, "SIR-based transmit power control of reverse link for coherent DS-CDMA mobile radio," IEICE Trans. Commun., vol. E81-B, pp. 1508-1516, July 1998.

[18] Y. Okumura and F. Adachi, "Variable rate data transmission with blind rate detection for coherent DS-CDMA mobile radio," IEE Electron. Lett., vol. 32, pp. 1865-1866, Sept. 1996.

[19] A. Shibutani, H. Suda, and F. Adachi, "Multi-stage interleaving method for turbo codes in DS-CDMA mobile radio," Proc. ICCS/APCC'98, pp. 391-395, Singapore, Nov. 23-27, 1998.

[20] K. Higuchi, M. Sawahashi, and F. Adachi, "Experiments on fast cell search algorithm for inter-cell asynchronous W-CDMA mobile radio," IEE Electron. Lett., vol. 35, pp. 1046-1047, June 1999.

[21] F. Kikuchi, H. Suda, and F. Adachi, "Effect of fast transmit power control on DS-CDMA forward link capacity," IEICE Trans. Commun., vol. E83-B, pp. 47-55, Jan. 2000.

[22] H. Suda, H. Kawai, and F. Adachi, "A fast transmit power control based on Markov transition for DS-CDMA mobile radio," IEICE Trans. Commun., vol. E82-B, pp. 1353-1362, Aug. 1999.

[23] C. Berrou, A. Glavieux, and P. Thitimajshima, "Near Shannon limit error-correcting coding and decoding: Turbo-codes," Proc. IEEE ICC '93, pp. 1064-1070, Geneva, Switzerland, May 1993.

[24] A. Fujiwara, H. Suda, and F. Adachi, "Application of turbo codes to DS-CDMA mobile radio," to be presented at IEEE Globecom'98, Sydney, Australia, 8-12 Nov. 1998.

[25] F. Adachi, M. Sawahashi, and H. Suda, "Promising techniques to enhance radio link performance of wideband wireless access based on DS-CDMA," IEICE Trans. Fundamentals, vol. E81-A, pp. 2242-2249, Nov. 1998.

[26] A. Duel-Hallen, J. Holtzman, and Z. Zvonar, "Multiuser detection for CDMA systems," IEEE Personal Commun., vol. 2, pp. 46-58, April 1995.

[27] M. K. Varanasi and B. A. Aazhang, "Multistage detection in asynchronous code-division multiple-access communications," IEEE Trans. Commun., vol. COM-38, pp. 509-519, April 1990.

[28] M. Sawahashi, Y. Miki, H. Andoh, and K. Higuchi, "Pilot symbol-aided coherent multistage interference canceller using recursive channel estimation for DS-CDMA mobile radio," IEICE Trans. Commun., vol. E79-B, pp. 1262-1270, Sept. 1996.

[29] M. Sawahashi, H. Andoh, and K. Higuchi, "Interference rejection weight control for pilot symbol-assisted coherent multistage interference canceller using recursive channel estimation in DS-CDMA mobile radio," IEICE Trans. Commun., vol. E81-A, pp. 957-972, May 1998.

[30] R.T. Compton Jr., "ADAPTIVE ANTENNAS: Concepts and Performance ," Prentice Hall, 1988.

[31] J. S. Thompson, P. M. Grant, and B. Mulgrew, "Smart antenna arrays for CDMA systems," IEEE Personal Commun. Mag., pp. 16-25, Oct. 1996.

[32] R. Kohno, "Spatial and temporal communication theory using adaptive antenna array," IEEE Personal Commun. Mag., pp.28-35, Feb. 1998.

[33] H. Wang, R. Kohno, and H. Imai, "Adaptive array Antenna combined with tapped delay line using processing gain for direct-sequence/spread-spectrum multiple access system," IEICE Trans Commun., vol. J75-B-II, pp. 815-825, Nov. 1992.

[34] S. Tanaka, M. Sawahashi, and F. Adachi, "Pilot symbol-assisted decision-directed coherent adaptive array diversity for DS-CDMA mobile radio reverse link," IEICE Trans. Fundamentals, vol. E80-A, pp. 2445-2454, Dec. 1997.

[35] M. Hata, "Future multimedia mobile systems beyond IMT-2000," Doc. MMC-6, APT Seminar on Multimedia Mobile Communications, 8-11 June 1999, Tokyo, Japan.

Abbreviations and Acronyms

Numerical

1G	First Generation
2G	Second Generation
3G	Third Generation
3G-GGSN	GGSN for 3G system
3G-SGSN	SGSN for 3G system
3G.IP	Third Generation All IP Network
3GPP	Third Generation Partnership Project

A

A/D	Analog-to-Digital
AAA	Authentication, Authorization and Accounting
AAC	Advanced Audio Coding
AAL	ATM Adaptation Layer
AC	1) Alternating Current
	2) Authentication Center
ACF	Association Control Function
ADC	A/D converter
ADPCM	Adaptive Differential Pulse Code Modulation
AGC	Automatic Gain Control
AHS	Automated Highway Systems
AMPS	Advanced Mobile Phone System
ANSI	American National Standards Institute
AP	Access Point
APN	Access Point Name
AR	Access Router
ARP	Address Resolution Protocol
ARIB	Association of Radio Industries and Businesses
ASIC	Application Specific Integrated Circuit
ASK	Amplitude Shift Keying
ATM	Asynchronous Transfer Mode
ATSC	Advanced Television Systems Committee
ATV	Advanced Television
AWGN	Additive White Gaussian Noise

B

BBU	Baseband Unit

BCJR	Bahl, Cocke, Jelinek and Raviv
BEP	Bit Error Probability
BER	Bit Error Rate
BICM	Bit-Interleaved Coded Modulation
BLAST	Bell Labs Layered Space-Time
BPF	Band Pass Filter
BPSK	Binary Phase Shift Keying
BRAN	Broadband Radio Access Networks
BS	1) Broadcast Satellite, 2) Base Station
BSC	Base Station Controller
BSS	1) Base Station System
	2) Basic Service Set
BST-OFDM	Band Segmented OFDM
BTS	Base Transceiver Station

C

CCD	Charge Coupled Device
CCI	Co-Channel Interference
CDMA	Code Division Multiple Access
CDPD	Cellular Digital Packet Data
CEIGU	Channel Estimation-and-Interference Replica Generation Unit
CFB-ROF	Common Frequency Band Radio on Fiber
CH	Correspondent Host
CL	Convergence Layer
CMA	Constant Modulus Algorithm
CMOS	Complementary Metal Oxide Semiconductor
COA	Care of Address
COFDM	Coded OFDM
COMSIC	Coherent Multi-Stage Interference Cancellation
CORBA	Common Object Request Broker Architecture
CPU	Central Processing Unit
CRC	Cyclic Redundancy Check
CRL	Communications Research Laboratory
CS	Control Station
CSI	Channel State Information
CSMA/CA	Carrier Sense Multiple Access with Collision Avoidance

D

D-GPS	Differential GPS
DAB	Digital Audio Broadcasting
DAT	Digital Audio Tape
dB	Decibel(s)
DC	Direct Current
DCC	DLC Connection Control
DDC	Digital Down Converter
DECT	Digital Enhanced Cordless Telecommunications
DEQPSK	Differentially Encoded QPSK
DFSK	Double Frequency Shift Keying
DFT	Discrete Fourier Transform
DHCP	Dynamic Host Configuration Protocol
DL	Downlink
DLC	Data Link Control Layer
DOA	Directions of Arrival
DOT	Directions of Transmission
DPSK	Differential Phase Shift Keying
DQPSK	Differential QPSK
DS	Direct Sequence
DS/CDMA (also DS-CDMA)	Direct Sequence/Code Division Multiple Access
DS/SS	Direct Sequence/Spread Spectrum
DSP	Digital Signal Processor
DSPH	Digital Signal Processing Hardware
DSPS	Digital Signal Processing Software
DTV	Digital Television
DVB	Digital Video Broadcasting
DVB-T	Digital Video Broadcasting - Terrestrial

E

EAM	Electroabsorption Modulator
EDGE	Enhanced Data Rates for GSM Evolution
EEP	Equal Error Protection
EGGSN	Enhanced Gateway GPRS Support Node
EGPRS	Enhanced GPRS
EPG	Electronic Program Guide
ESGSN	Enhanced Serving GPRS Support Node
ESPRIT	Estimation of Signal Parameters via Rotational Techniques
ESS	Extended Service Set

ETC	Electronic Tall Collection
ETSI	European Telecommunication Standards Institute
ETSI SMG	Special Mobile Group in ETSI
EU	European Union
EVM	Error Vector Magnitude
E_b/I_0	Energy per Information Bit to Interference plus Background Noise Spectral Density
E_b/N_0	Energy per Bit to Noise Spectral Density

F

FA	Foreign Agent
FC	Frequency Converter
FDD	Frequency Division Duplex
FDMA	Frequency Division Multiple Access
FDP	Frequency Domain Pilot
FEC	Forward Error Correction
FER	Frame Error Rate
FFT	Fast Fourier Transform
FH	Frequency Hopping
FIR	Finite Impulse Response
FL	Forward Link
FM	Frequency Modulation
FN	Foreign Network
FPGA	Field Programmable Gate Array
FPLMTS	Future Public Land Mobile Telecommunication Systems
FSK	Frequency Shift Keying
FWA	Fixed Wireless Access

G

G3G	Global 3G
GGSN	Gateway GPRS Supported Node
GI	1) Group Identification, 2) Guard Interval
GII	Guard Interval Insertion
GMSC	Gateway Mobile Switching Center
GMSK	Gaussian-filtered Minimum Shift Keying
GPRS	General Packet Radio Service
GPS	Global Positioning System
GR	Gateway Router
GSIC	Groupwise Serial Interference Cancellation

GSM	Global Systems for Mobile Communications
GSM-MAP	GSM Mobile Application Part
GTP	GPRS Tunneling Protocol
GaAs	Gallium Arsenide

H

HA	Home Agent
HD	Hard Decision
HD-PIC	Parallel Interference Cancellation with Hard Decision
HDTV	High Definition Television
HF	High Frequency
HIPERLAN	High Performance Radio Local Area Network
HIPERLAN/1	HIPERLAN Type 1
HIPERLAN/2	HIPERLAN Type 2
HLR	Home Location Register
HN	Home Network
HTML	HyperText Markup Language

I

IAPP	Inter Access Point Protocol
IC	Interference Cancellation
Ich	In-Phase Channel
ICMP	Internet Control Message Protocol
IDFT	Inverse Discrete Fourier Transform
IDL	Interface Definition Language
IEEE	Institute for Electrical and Electronics Engineers
IETF	Internet Engineering Task Force
IF	Intermediate Frequency
IFU	IF Unit
i.i.d.	independent, identically distributed
IIR	Infinite Impulse Response
IMT-2000	International Mobile Telecommunications 2000
IP	1) Internet Protocol
	2) Information Provider
IPI	Interpath Interference
IPv4	Internet Protocol version 4
IPv6	Internet Protocol version 6

IS-136	Interim Standard 136
IS-54	Interim Standard 54
IS-95	Interim Standard 95
ISDB	Integrated Services Digital Broadcasting
ISDB-C	ISDB - Cable
ISDB-S	ISDB - Satellite
ISDB-T	ISDB - Terrestrial
ISI	Intersymbol Interference
ISM	Industrial, Scientific And Medical
ITS	Intelligent Transport System
ITU	International Telecommunication Union
ITU-R	International Telecommunication Union Radiocommunication Sector
IVC	Inter-Vehicle Communications

L

LAN	Local Area Network
LBS	Local Base Station
LD	Laser Diode
LDTV	Low Definition Television
LLC	Logical Link Control
LMMSE	Linear Minimum Mean Square Error
LMS	Least Mean Square
LO	Local Oscillator
LPF	Low Pass Filter
LUT	Look Up Table

M

MAC	1) Medium Access Control 2) Multiply and Accumulate
MAI	Multiple-Access Interference
MCM	Multi-Carrier Modulation
MEM	Maximum Entropy Method
MExE	ETSI's Mobile Station Application Exchange Environment
MF	Matched Filter
MFB	Matched Filter Bound
MFN	Multi Frequency Network
MH	Mobile Host
MIL	Multi-stage Interleaver
MIPS	Million Instructions per Second
ML	Maximum Likelihood

MLS	Multimedia Lane and Station
MLSD	Maximum Likelihood Sequence Detector
MMIC	Monolithic, Microwave Integrated Circuit
MMSE	Minimum Mean Square Error
MNRP	Mobile Network Registration Protocol
MOTIVATIVE	Mobile Television and Innovative Receiver
MPEG	Moving Picture Experts Group
MPT	Ministry of Posts and Telecommunications
MRC	Maximal Ratio Combining
MS	1) Mobile Station, 2) Mobile Subscriber
MSC	Mobile Switching Center
MT	Mobile Terminal
MUD	Multiuser Detection
MUSIC	Multiple Signal Classification

N

NTD	Network Terminating Devices
NTSC	National Television System Committee
NTT	Nippon Telegraph and Telephone
NW	Network

O

OFDM	Orthogonal Frequency Division Multiplexing
OSI	Open Systems Interconnection
OVSF	Orthogonal Variable Spreading Factor

P

P/S	Parallel-to-Serial
PAM	Pulse Amplitude Modulation
PCF	Packet Control Function
PCS	Personal Communication Services
PD	Photo Detector
PDC	Personal Digital Cellular
PDC-P	PDC-Packet
PDCP	Packet Data Convergence Protocol
pdf	probability density function
PDP	Packet Data Protocol
PDSN	Packet Data Serving Node
PDTCH	Packet Data Traffic Channel

PDU	Packet Data Unit
PHS	Personal Handy-phone System
PHY	Physical Layer
PIC	Parallel Interference Cancellation
PN	Pseudo Noise
PPP	Point-to-Point Protocol
PRBS	Pseudo-Random Binary Sequence
PRS	Partial Response Signaling
PSK	Phase Shift Keying
PSTN	Public Switched Telephone Network

Q

QAM	Quadrature Amplitude Modulation
Qch	Quadrature-Phase Channel
QoS	Quality of Services
QPSK	Quaternary Phase Shift Keying

R

RAC	Radio Access Control
RADIUS	Remote Authentication Digital In User Service
RAN	Radio Access Network
RARP	Reverse ARP
RF	Radio Frequency
RFU	RF Unit
RL	Reverse Link
RLC	Radio Link Control
RLP	Radio Link Protocol
RLS	Recursive Least Squres
RMS	Root Mean Square
RNC	Radio Network Controller
ROF	Radio on Fiber
RRC	Radio Resource Control
RS	Reed Solomon
RSC	Recursive Systematic Code
RVC	Road-Vehicle Communications
RX	Receiver

S

S-WMF	Spatially Whitened Matched Filter
S/P	Serial-to-Parallel
SD	Soft Decision

SD-IC	Interference Cancellation with Soft Decision
SD-PIC	Parallel Interference Cancellation With Soft Decision
SDMA	Space Division Multiple Access
SDR	Software Defined Radio
SDTV	Standard Definition Television
SF	Spreading Factor
SFN	Single Frequency Network
SGSN	Serving GPRS Support Node
Si	Silicon
SIC	Serial Interference Cancellation
SIG	Special Interest Group
SIR	Signal-to-Interference Power Ratio
SNDCP	Sub-network Dependent Convergence Protocol
SNR	Signal to Noise Power Ratio
SS7	Signaling System No. 7
ST-WMF	Spatially and Temporally Whitened Matched Filter
ST-TF	Spatial and Temporal Transmission Filter
STBC	Space Time Block Code
STC	Space-Time Coding
SWAP	Shared Wireless Access Protocol

T

T-WMF	Temporally Whitened Matched Filter
TACS	Total Access Communication System
TBF	Temporary Block Flow
TC-8PSK	Trellis Coded 8PSK
TCM	Trellis Coded Modulation
TCP/IP	Transmission Control Protocol and Internet Protocol
TD-CDMA	Hybrid TDMA/CDMA system
TDD	Time Division Duplex
TDL	Tapped Delay Line
TDMA	Time Division Multiple Access
TDP	Time Domain Pilot
TFI	Temporary Flow Identifier
TFT	Traffic Flow Template
TG8/1	Task Group 8/1
TIA	Telecommunications Industry Association

TMCC	Transmission and Multiplexing Configuration Control
TPC	Transmit Power Control
TX	Transmitter

U

U-NII	Unlicensed National Information Infrastructure
UDP	User Datagram Protocol
UEP	Unequal Error Protection
UL	Uplink
UML	Unified Modeling Language
UMTS	Universal Mobile Telecommunication Systems
USB	Universal Serial Bus
USF	Uplink State Flag
UTRA	UMTS Terrestrial Radio Access
UTRAN	UMTS Terrestrial Radio Access Network
UWC-136	Universal Wireless Communication 136

V

VD	1) Viterbi Decoder, 2) Viterbi Detector
VICS	Vehicle Information and Communication Systems
VLR	Visitor Location Register
VSB	Vestigial Side Band
VoIP	Voice over IP

W

W-CDMA	Wideband CDMA
WAP	Wireless Application Protocol
WARC	World Administrative Radio Conference
WLAN	Wireless LAN
WMF	Whitened Matched Filter
WMSA	Weighted Multi-Slot Averaging
WRC	World Radiocommunication Conference

Y

| YRP | Yokosuka Research Park |

Z

ZF   Zero-Forcing

Index

About the Editors

NORIHIKO MORINAGA received the B.E. degree in electrical engineering from Shizuoka University, Shizuoka, Japan, in 1963 and the M.E. and Ph. D. degrees from Osaka University, Osaka, Japan in 1965 and 1968, respectively. He is currently a Professor in the Department of Communications Engineering, Osaka University, working in the area of radio, mobile, satellite, and optical communication systems and EMC. He was the General Chairman of the 10th International Symposium of Personal, Indoor and Mobile Radio Communications (PIMRC'99) held in Osaka during 12-15 September 1999, and he is a Steering Bord Member of the International Symposium on Wireless Personal Multimedia Communications (WPMC) since 1998. At present, he is the Vice President of Institute of Electronics, Information and Communication Engineers. He was the President of the Communications Society of IEICE (1998-1999), President-Elect of the Communications Society of IEICE (1997-1998), Editor-in-Chief of the Transactions of IEICE B (1995-1997), the Chairman of the Satellite Telecommunications Technical Group of IEICE (1990-1991) and the Chairman of the Radio Communication Systems Technical Group (1989-1990) of IEICE. Prof. Morinaga received the Telecom Natural Science Award (1986), and the Telecom System Technology Award (1994) from the Telecommunication Advancement Foundation, and the Paper Award from the IEICE (1996). He is a senior member of the IEEE, a member of IEICE and the Institute of Image Information and Television Engineers.

RYUJI KOHNO was born in Kyoto, Japan March 1956. He received the B.E. and M.E. degrees in computer engineering from Yokohama National University in 1979 and 1981, respectively, and the Ph.D. degree in electrical engineering from the University of Tokyo in 1984. He joined in the Department of Electrical Engineering, Toyo University in 1984 and became an Associate Professor in 1986. During 1988-1997 he was an Associate Professor in the Division of Electrical and Computer Engineering, Yokohama National University. Since 1998 he is a Professor in the same division. During 1984-1985 he was a Visiting Scientist in the Department of Electrical Engineering, the University of Toronto. At the present, he is the Chairman of both the Society of Intelligent Transport System (ITS) and the Society of Software Radio of the IEICE. He is currently an Editor of the IEEE Transactions on Communications and that of the IEICE Transactions on Fundamentals of Electronics, Communications, and Computer Sciences. He also plays a role of a Director of the Society of Information Theory and its Applications and so on. He has been elected a member of the Board of Governors of IEEE Information Theory Society for a three-year term beginning 1 January 2000. He was an editor of the IEICE Transactions on Communications (English Volume) for six years and an associate editor of the IEEE Transactions on Information Theory for four years (1995-1998). He was the Chairman of the society of Spread Spectrum Technology of the IEICE (1995-1998), the Chairman of the Technical Program Committee (TPC) of 1992 IEEE International Symposium on Spread-Spectrum Techniques and Applications (ISSSTA'92), the TPC Vice-Chairman of 1993 International Symposium on Personal Indoor and Mobile Radio Communications (PIMRC'93), an executive organizer of 1993 IEEE International Workshop on Information theory (ITW'93), the TPC Co-chairman of 1996 IEEE International Workshop on Intelligent Signal Processing & Communication Systems, the TPC Chairman of 1999 IEEE International Symposium on Personal, Indoor, Mobile Radio Communications (PIMRC'99), the TPC Chair of 2000 International Workshop on Personal and Mobile Radio Communications (WPMC'2000) and so on. His current research interests lie in the areas of space-time signal processing, coding theory, spread spectrum system, array antenna, software radio, and their applications to various kinds of practical communication systems and intelligent transport systems (ITS). He is a member of IEEE, EURASIP, IEICE, IEE of Japan, IPS of Japan. He wrote technical books entitled *Spread Spectrum Techniques and Applications*, *Digital Signal Processing*, *Data Communication Systems* and is currently writing the books entitled *Advanced Spread Spectrum Techniques and Applications* and *Smart Antenna: Spatial and Temporal Communication Theory* etc.

SEIICHI SAMPEI received the B.E., M.E. and Ph.D. degrees in electrical engineering from Tokyo Institute of Technology in 1980, 1982 and 1991, respectively. From 1982 to 1993, he was with Communications Research Laboratory (CRL), Ministry of Posts and Telecommunications, Japan, where he was engaged in developing adjacent channel interference rejection, fading compensation and M-ary QAM technologies for wireless communication systems. From 1991 to 1992, he was at the University of California, Davis as a visiting researcher. In 1993, he joined the Department of Communications Engineering, Osaka University as an Associate Professor, where he is currently developing intelligent transmission and access technologies for wireless communication systems that include adaptive modulation and adaptive access control technologies. When he was with the CRL, he was a member of digital MCA System standardization committee in ARIB and a Japanese delegate of ITU-R TG8/1. He is currently a special member of IMT-2000 Committee in ARIB. He was the Secretary of PIMRC'99, and an Executive Committee Member of the 4th Asia-Pacific Conference on Communication (APCC'95) and PIMRC'93. He is a TPC member of VTC-2000-SPRING. He authored technical books entitled *Applications of Digital Wireless Technologies to Global Wireless Communications* (Prentice-Hall 1997), and *Wireless Multimedia Network Technologies* (Kluwer 1999). He received the Shinohara Young Engineering Award from the

IEICE (1986) and the Telecom System Technology Award from the Telecommunication Advancement Foundation (1992). He is a member of IEICE, IEEE and the Institute of Image Information and Television Engineers.

Contributors

Part I: NEW TECHNICAL TREND IN WIRELESS MULTIMEDIA COM-MUNICATIONS

Chapter 1 Spatial Channel Modeling for Wireless Communications

GREGORY D. DURGIN was born in Baltimore, Maryland, on October 23, 1974. He received the B.S.E.E. and M.S.E.E. from Virginia Tech in 1996 and 1998, respectively. He is currently a Bradley Fellow at Virginia Tech, working toward the Ph.D. degree at the Mobile & Portable Radio Research Group (MPRG). Since 1996, he has been a research assistant at MPRG, where his research focuses on radio wave propagation, channel measurement, and applied electromagnetics. He received the 1998 Blackwell Award for best graduate research presentation in the electrical and computer engineering department at Virginia Tech. He received the 1999 Stephen O. Rice Prize, with co-authors Theodore S. Rappaport and Hao Xu, for best original research paper published in the *IEEE Transactions on Communications*. As a student, he has published 14 technical papers in international journals and conferences. He also serves regularly as a consultant to industry.

THEODORE S. RAPPAPORT received BSEE, MSEE, and Ph.D. degrees from Purdue University in 1982, 1984, and 1987, respectively. Since 1988, he has been on the Virginia Tech electrical and computer engineering faculty, where he is the James S. Tucker Professor and founding director of the Mobile & Portable Radio Research Group (MPRG), a university research and teaching center dedicated to the wireless communications field. In 1989, he founded TSR Technologies, Inc., a cellular radio/PCS manufacturing firm that he sold in 1993. He received the Marconi Young Scientist Award in 1990 and an NSF Presidential Faculty Fellowship in 1992. Dr. Rappaport holds 3 patents and has authored, co-authored and co-edited 14 books in the wireless field, including the popular textbook *Wireless Communications: Principles & Practice* (Prentice-Hall, 1996), *Smart Antennas for Wireless Communications: IS-95 and Third Generation CDMA Applications* (Prentice Hall, 1999), and several compendia of papers, including *Cellular Radio & Personal Communications: Selected Readings* (IEEE Press, 1995), *Cellular Radio & Personal Communications: Advanced Selected Readings* (IEEE Press, 1996), and *Smart Antennas: Selected Readings* (IEEE Press, 1998). He has co-authored more than 130 technical journal and conference papers and was recipient of the 1998 IEEE Communications Society Stephen O. Rice Prize Paper Award. He serves on the editorial board of *International Journal of Wireless Information Networks* (Plenum Press, NY), is a Fellow of the IEEE, and is active in the IEEE Communications and Vehicular Technology societies. Dr. Rappaport is also chairman of Wireless Valley Communications, Inc., a microcell and in-building design and development firm. He is a registered professional engineer in the state of Virginia and is a Fellow and past member of the board of directors of the Radio Club of America. He has consulted for over 20 multinational corporations and has served the International Telecommunications Union as a consultant for emerging nations.

Chapter 2 Space-Time Coding and Signal Processing for High Data Rate Wireless Communications

AYMAN NAGUIB received the B.Sc degree (with honors) and the M.S. degree in electrical engineering from Cairo University, Cairo, Egypt, in 1987 and 1990 respectively, and the M.S. degree in statistics and the Ph.D. degree in electrical engineering from Stanford University, Stanford, CA, in 1993 and 1996, respectively. From 1987 to 1989, he spent his military service at the Signal Processing Laboratory, The Military Technical College, Cairo, Egypt. From 1989 to 1990, he was employed with Cairo University as a research and teaching assistant in the Communication Theory Group, Department of Electrical Engineering. From 1990 to 1995, he was a research and teaching assistant in the Information Systems Laboratory, Stanford University, Stanford, CA. In 1996, he joined AT & T Labs, Florham Park, NJ, where he is now a principal member of technical staff. His current research interests include antenna arrays, signal processing, modulation, and coding for high data rate wireless and digital communications and modem design for broadband systems.

A. ROBERT CALDERBANK received the B.S. degree in 1975 from Warwick University, U.K., the M.S. degree in 1976 from Oxford University, U.K., and the Ph.D. degree in 1980 from California Institute of Technology, Pasadena, all in Mathematics. He joined AT & T Bell Laboratories in 1980, and prior to the split of AT & T and Lucent, he was a Department Head in the Mathematical Sciences Research Center at Murray Hill. He is now Director of the Information Sciences Research Center at AT & T Labs - Research in Florham Park, NJ. His research interests range from algebraic coding theory to wireless data transmission to quantum computing. At the University of Michigan and at Princeton University, he has developed and taught an innovative course on bandwidth-efficient communication. From 1986 to 1989, Dr. Calderbank was Associate Editor for Coding Techniques for the IEEE Transactions on Information Theory. From 1996 to 1999, he was the Editor-in-Chief of the IEEE transactions on Information theory. He was also Guest Editor for the Special Issue on of the *IEEE Transactions on Information Theory* dedicated to coding for storage devices. He served on the board of Governors of the IEEE Information Theory Society from 1990 to 1996. Dr. Calderbank received the 1995 Prize Paper Award from the Information Theory Society for his work on the Z4 linearity of the Kerdock and Preparata codes (jointly with A. R. Hammons, Jr., P. V. Kumar, N. J. A. Sloane and P Sole). He also received the 1999 Information Theory Society Best Paper Award (jointly with V. Tarokh and N. Seshadri).

Chapter 3 Coding for the Wireless Channel

EZIO BIGLIERI was born in Aosta (Italy) in 1944. He received his training in Electrical Engineering from Politecnico di Torino (Italy), where he received his Dr. Engr. degree (*summa cum laude*) in 1967. From 1968 to 1975 he was with the Istituto di Elettronica e Telecomunicazioni, Politecnico di Torino, first as a Research Engineer, then as an Associate Professor (jointly with Istituto Matematico). In 1975 he was made a Full Professor of Electrical Engineering at the University of Napoli (Italy). In 1977 he returned to Politecnico di Torino as a Professor in the Department of Electrical Engineering. From 1987 to 1990 he was a Professor of Electrical Engineering at the University of California, Los Angeles. Since 1990 he has been again a professor with Politecnico di Torino. From January to June 1977 he was a Visiting Lecturer and Research Engineer in the Department of System Science, UCLA. He spent the summers of 1979 and 1982 working with the Mathematical Research Center, Bell Laboratories, Murray Hill, NJ, and with the Bell Laboratories, Holmdel, NJ, respectively. In May–June 1984 he was a Visiting Research Engineer with the Department of Electrical Engineering, UCLA, and in the Spring of 1986 and of 1999 he was a Visiting Professor in the same Department. In February–September 1994 he was a Visiting Professor with the Ecole Nationale Supérieure de Télécommunications, Paris, France. In August 1997 he was a Visiting Professor at the University of Sydney, Australia. In October 1998–January 1999 he was a Visiting Professor at Yokohama National University, Japan. From April to August 1999 he was a Visiting Professor with the Department of Electrical Engineering, UCLA, and from February to May 2000 a Visiting Professor with Princeton University. In 1996–1997 he served as chairman of the IEEE Communications Society Awards Committee. In 1988, 1992, and 1996 he was elected to the Board of Governors of the IEEE Information Theory Society. In 1999 he was the President of the Society, and he is actually serving as its Past President. He is the general co-chairman of the "IEEE 2000 International Symposium on Information Theory," Sorrento, Italy, June 2000. From 1988 to 1991 he was an Editor of the *IEEE Transactions on Communications*, and from 1991 to 1994 an Associate Editor of the *IEEE Transactions on Information Theory*. From 1997 to 1999 he was an Editor of the *IEEE Communications Letters*, and, since 1998, he has been a Division Editor of the *Journal of Communications and Networking*. From 1991 to 1997 he was an Editor of the *European Transactions on Telecommunications*. He is now the Editor in Chief of this journal. In 1992 he received the "IEE Benefactors Premium" from the Institution of Electrical Engineers (U.K.) for a paper on trellis-coded modulation. In 2000 he received the "IEEE Donald G. Fink Prize Paper Award," presented for the most outstanding survey, review, or tutorial paper published by IEEE in 1999. In 2000 he also received the IEEE Third-Millennium Medal for outstanding contributions to the Information Theory area of technology. He is a *Fellow* of the IEEE.

GIORGIO TARICCO was born in Torino (Italy) in 1961. He received his training in Electrical Engineering from Politecnico di Torino (Italy), where he received his Dr. Engr. degree in 1985. From 1985 to 1987 he was with CSELT (Italian Telecom Labs) where he was involved in the study and definition of the GSM communication system with special regard to the performance of the channel coding subsystem. Since 1991 to 1994 he was with the Dipartimento di Elettronica of Politecnico di Torino as a Researcher. Since 1995, still with the Dipartimento di Elettronica, he has been a Professor of Analog and Digital Communications. Since 1993 he has been involved in several ESTEC contracts with Politecnico di Torino and in Summer 1996 he was a Research Fellow at ESTEC. Prof. Taricco is a member of the IEEE. He was a Session Chair in several IEEE conferences and he is the Finance Chairman of IEEE ISIT 2000. His research interests are in the areas of error-control coding, digital communications, multiuser detection and information theory with application to mobile communication systems.

GUISEPPE CAIRE was born in Torino, Italy, on May 21, 1965. He received the B.Sc. in Electrical Engineering from Politecnico di Torino (Italy), in 1990, the M.Sc. in Electrical Engineering from Princeton University (USA) in 1992 and the Ph.D. from Politecnico di Torino in 1994. He was a recipient of the AEI G.Someda Scholarship in 1991, has been with the European Space Agency (ESTEC, Noordwijk, The Netherlands) in 1995, was a recipient of the COTRAO Scholarship in 1996 and a CNR Scholarship in 1997. He has been visiting the Institute Eurecom, Sophia Antipolis, France, in 1996 and Princeton University in summer 1997. He has been Assistant Professor in Telecommunications at the Politecnico di Torino and presently he is Associate Professor with the Department of Mobile Communications of Eurecom Institute and Associate Editor for CDMA and Multiuser Detection of the *IEEE Transactions on Communications*. He is co-author of more than 30 papers in international journals and more than 50 in international conferences, and he is author of three international patents with the European Space Agency. His interests are focused on digital communications theory, information theory, coding theory and multiuser detection, with particular focus on wireless terrestrial and satellite applications.

Chapter 4 OFDM: The Most Elegant Solution for Wireless Digital Transmission

SHINSUKE HARA received the B.Eng., M.Eng. and Ph. D. degrees in communication engineering from Osaka University, Osaka, Japan, in 1985, 1987 and 1990, respectively. From April 1990 to March 1996, he was an Assistant Professor in the Department of Communication Engineering, Osaka University. Since April 1996, he has been with the Department of Electronic, Information and Energy Engineering, Graduate School of Engineering, Osaka University, and now, he is an Associate Professor. Also from April 1995 to March 1996, he was a Visiting Scientist at Telecommunications and Traffic Control Systems Group, Delft University of Technology, Delft, The Netherlands. His research interests include satellite, mobile and indoor wireless communications systems, and digital signal processing.

Chapter 5 Overview of Linear Multiuser Equalizers for DS/CDMA Systems

MARKKU J. JUNTTI was born in Kemi, Finland, in 1969. He received his M.Sc. (Tech.) and Dr.Sc. (Tech.) degrees in Electrical Engineering from University of Oulu, Oulu, Finland in 1993 and 1997, respectively. Dr. Juntti has been a Research Scientist and Research Project Manager at Telecommunication Laboratory and Centre for Wireless Communications, University of Oulu in 1992–97. In academic year 1994–95 he was a Visiting Research Scientist at Rice University, Houston, Texas. In 1998 he was an Acting Professor at the University of Oulu. In 1999–2000 he was with Nokia Networks, Radio Access Systems in Oulu as a Senior Specialist. Dr. Juntti has been a Professor of Telecommunications at University of Oulu since 2000. Dr. Juntti's research interests include communication theory and signal processing for wireless communication systems as well as their application in wireless communication system design. Dr. Juntti is a member of IEEE. He was Secretary of IEEE Communication Society Finland Chapter in 1996–97 and has been elected the Chairman for years 2000-01. He has been Chairman of Technical Program Committees of 1999 Finnish Signal Processing Symposium (FINSIG'99) and the 2000 Finnish Workshop on Wireless Communications (FWCW'00).

KARI J. HOOLI was born in Espoo, Finland, in 1972. He received his M.Sc. (Tech.) degree in Electrical Engineering from University of Oulu, Oulu, Finland in 1998. K. Hooli has been a Research Scientist at Centre for Wireless

Communications, University of Oulu since 1997. Hooli's research interests include signal processing for wireless communication systems with emphasis on detection and equalization. Hooli is a student member of IEEE.

Chapter 6 Software-Defined Radio Technologies

SHINICHIRO HARUYAMA is a researcher at Advanced Telecommunication Laboratory of SONY Computer Science Laboratories, Inc., Tokyo, Japan. He also serves as visiting associate professor at Keio University, Yokohama, Japan. He received an M. S. in engineering science from University of California at Berkeley in U.S.A. and a Ph.D. in computer science from the University of Texas at Austin in U.S.A. Since 1991, he has worked at Bell Laboratories of Lucent Technologies, U.S.A. before joining Keio University in 1997. He joined SONY Computer Science Laboratories, Inc. in 1998. His research interests include software radio, wireless communication, reconfigurable system, FPGA, and VLSI design automation.

Chapter 7 Spatial and Temporal Communication Theory based on Adaptive Antenna Array

RYUJI KOHNO: (See About the Editors)

Part II: TRENDS IN NEW WIRELESS MULTIMEDIA COMMUNICATION SYSTEMS

Chapter 8 Intelligent Transport Systems

MASAYUKI FUJISE was born in Fukuoka, Japan, on December 8, 1950. He received the B.S., M.S. and Dr. Eng. degrees, in communication engineering from Kyushu University, Fukuoka, Japan, in 1973, 1975 and 1987, respectively and the M. Eng. degree in electrical engineering from Cornell University, Ithaca, NY, in 1980. He joined KDD in 1975 and was with the R & D Laboratories being engaged in research on optical fiber measurement technologies for optical fiber transmission systems. In 1990, he joined ATR Optical and Radio Communications Research Laboratories as a department head, where he managed research on optical inter-satellite communications and active array antenna for mobile satellite communications. Since he joined Communications Research Laboratory Ministry of Posts and Telecommunications in 1997, he has been an executive manager of millimeter-wave applications group for Intelligent Transport Systems. He is now interested in radio on fiber technology and software defined radio technology etc. Dr. Fujise is the recipient of the Jack Spergel Memorial Award of the 33rd International Wire & Cable Symposium in 1984 and he is a member of the IEICE and the IEEE.

AKIHITO KATO was born in Osaka Japan, in 1965. He received the B.E. degree of electrical engineering from Doshisha University, Japan in 1989 and he also received D.E. degree of electrical engineering from Doshisha University in 1994. He joined the Communications Research Laboratory, Ministry of Posts and Telecommunications, Tokyo, in 1994. Since then, he has been engaged in research on millimeter-wave indoor propagation and millimeter-wave wireless communication systems such as ultra high-speed wireless LAN system and inter-vehicle communication system on ITS. He is currently a senior researcher of Yokosuka Radio Communications Research Center of the CRL. Dr. Kato is a member of IEICE, Japan and IEEE.

KATSUYOSHI SATO was born in Iwate, Japan, in 1967. He received the B.S. and M.S. degrees in electronic engineering from Tohoku University, Sendai, Japan, in 1989 and 1991, respectively. In 1991, he joined the communications Research Laboratory, Ministry of Posts and Telecommunications, Tokyo, Japan. He has been engaged in research on radio propagation characteristics of millimeter-wave, such as indoor radio propagation and millimeter-wave remote sensing. His current research interests include wireless technologies for ITS. He is a member of the Institute of Electronics, Information and Communication Engineers (IEICE) of Japan and the Physical Society of Japan.

HIROSHI HARADA was born in Kobe, Japan, in 1969. He received M.E. and Ph.D degrees from Osaka university, Osaka, Japan in 1994 and 1995 respectively. From 1995, he joined the Communications Research Laboratory (CRL), Ministry of Posts and Telecommunications (MPT), Japan, where he was involved in the areas of high speed mobile radio transmission techniques by using parallel transmission, e.g. multi-code and multi-carrier based transmission. From 1996 to 1997, he was a postdoctoral fellow of Delft University of Technology, The Netherlands, where he was engaged in the research of OFDM based mobile communication systems, especially radio transmission techniques. He is currently a researcher of CRL, MPT, Japan. His current research interests include digital-signal-processing based mobile telecommunication systems, e.g. software radio and multimedia mobile access communication (MMAC) systems. He received the Young Engineer Award of Institute of Electronics , Information and Communication Engineers (IEICE) of Japan in 1999. Dr. Harada is a member of the IEICE and IEEE.

Chapter 9 Wireless Data Communication Systems

KAVEH PAHLAVAN is a professor of ECE and Director of the Center for Wireless Information Network Studies, Worcester Polytechnic Institute, Worcester, MA. His area of research is broadband wireless local networks. His previous research background is on modulation, coding and adaptive signal processing for digital communication. He has contributed to more than 200 technical papers and presentations in various countries. He is the principal author of the Wireless Information Networks, John Wiley and Sons, 1995. He has been a consultant to many industries including CNR Inc., GTE Laboratories, Steinbrecher Corp., Simplex, Mercurry Computers, WINDATA, SieraComm, and Codex/Motorola in Massachusetts; JPL, Savi Technologies, RadioLAN in California, Airnoet in Ohio, Honeywell in Arizona; Nokia, LK-Products, Elektrobit, and TEKES in Finland, and NTT in Japan. Before joining WPI, he was the Director of Advanced Development at Infinite Inc., Andover, Mass. working on data communications. He started his career as an Assistant Professor at Northeastern University, Boston, MA. He is the Editor-in-Chief of the *International Journal on Wireless Information Networks*. He was the program chairman and organizer of the IEEE Wireless LAN Workshop, Worcester, in 1991 and 1996 and the organizer and the technical program chairman of the IEEE International Symposium on Personal, Indoor, and Mobile Radio Communications, Boston, MA, 1992 and 1998. For his contributions to the wireless networks he was the Westin Hadden Professor of Electrical and Computer Engineering at WPI during 1993-1996, and was elected as a Fellow of the IEEE in 1996 and become a Fellow of Nokia in 1999.

XINRONG LI is a Research Assistant at the Center for Wireless Information Network Studies, Worcester Polytechnic Institute, Worcester, MA. His recent research has focused on indoor geolocation techniques and issues in the fourth generation wireless data communication systems.

MIKA YLIANTTILA received his M.Sc. (E.E.) degree from the University of Oulu, Finland, in 1998. He is currently working as a project manager in the Centre for Wireless Communications at the University of Oulu, and he is working for Ph.D. degree in the area of mobility management and system architecture issues in the fourth generation wireless networks. His professional interest include IP protocol evolution, wireless optimizations, location based services and real-time architectures.

MATTI LATVA-AHO received M.Sc. (E.E.), Lic.Tech., and Dr. Tech. degrees from the University of Oulu, Finland in 1992, 1996 and 1998, respectively. From 1992 to 1993, he was a Research Engineer at Nokia Mobile Phones, Oulu, Finland. During the years 1994 - 1998 he was a Research Scientist at Telecommunication Laboratory and Centre for Wireless Communications at the University of Oulu. Currently Professor Latva-aho is Director of the Centre for Wireless Communications at the University of Oulu.

Chapter 10 Wireless Internet – Networking Aspect –

LI FUNG CHANG received the B.S. degree from the National Taiwan Normal University in 1978 and the M.S., Ph.D. degrees from the University of Illinois in 1983 and 1985, respectively. She has been with the broadband wireless systems research department, AT & T Labs Research since Feb. 1999. Prior to joining AT & T, she spent 13 years at Bellcore wireless research department where she was the director of the broadband wireless networking research group and was project manager for several government funded research works on tactical wireless communications. Her current research interests are in the area of wireless networking including wireless access to the Internet and system designs to support high speed wireless packet data communications. She holds eight US patents and several international patents, several pending patents in the area of wireless communications and has numerous publications. She has given several short courses in PCS related topics, such as "Air Interface Standards, Wireless Data Communications, Wireless ATM and Wireless Internet: Networking Aspect. She was a guest editor of the JSAC special issues on Wireless ATM, and is now one of the editors for the JSAC: Wireless Communication series. She is a senior member of IEEE, Phi Kappa Phi and Phi Tau Phi Chinese honor society.

Chapter 11 Digital Terrestrial TV Broadcasting Systems

MAKOTO ITAMI was born in 1961 in Japan. He received Ph.D. degree in electrical engineering from the University of Tokyo in 1989. From 1989, he has been working at Science University of Tokyo, where he is an assistant professor of the department. His research interests are in the area of communication systems and signal processing. Especially he is interested in spread spectrum communication, OFDM and intelligent transportation systems (ITS).

Chapter 12 IMT-2000 – Challenges of Wireless Millennium

FUMIYUKI ADACHI received his B.S. and Dr. Eng. degrees in electrical engineering from Tohoku University, Sendai, Japan, in 1973 and 1984, respectively. In April 1973, he joined the Electrical Communications Laboratories of Nippon Telegraph & Telephone Corporation (now, NTT) and conducted various research related to digital cellular mobile communications. From July 1992 to December 1999, he was with NTT Mobile Communications Network, Inc., where he led a research group on wideband/broadband CDMA wireless access for IMT-2000 and beyond. Since January 2000, he has been at Tohoku University, Sendai, Japan, where he is a Professor in the Department of Electrical Communications at Graduate School of Engineering. His research interests are in CDMA and TDMA wireless access techniques, CDMA spreading code design, Rake receiver, transmit/receive antenna diversity, adaptive antenna array, bandwidth-efficient digital modulation, and channel coding, with particular application to broadband wireless communications systems. From October 1984 to September 1985, he was a United Kingdom SERC Visiting Research Fellow in the Department of Electrical Engineering and Electronics at Liverpool University. From April 1997 to March 2000, he was a visiting Professor at Nara Institute of Science and Technology, Japan. He has published over 150 papers in journals and over 60 papers in international conferences. Dr. Adachi served as a Guest Editor of IEEE JSAC for special issue on Broadband Wireless Techniques, October 1999. He was a co-recipient of the IEEE Vehicular Technology Transactions Best Paper of the Year Award 1980 and again 1990. He is a member of Institute of Electronics, Information and Communication Engineers of Japan (IEICE) and was a co-recipient of the IEICE Transactions Best Paper of the Year Award 1996 and again 1998.

MAMORU SAWAHASHI received the B.S. and M.S. degrees from Tokyo University in 1983 and 1985, respectively, and received the Dr. Eng. Degree from Nara Institute of Science and Technology in 1998. In 1985 he joined NTT Electrical Communications Laboratories, and in 1992 he transferred to NTT Mobile Communications Network, Inc. Since joining NTT, he has been engaged in the research of modulation/demodulation techniques for mobile radio and research and development of radio transmission technologies for wideband DS-CDMA mobile radio. He is now a Executive Research Engineer in the Wireless Laboratories of NTT DoCoMo Inc.